Oxidants and Antioxidants in Cutaneous Biology

Current Problems in Dermatology

Vol. 29

Series Editor *G. Burg*, Zürich

Basel · Freiburg · Paris · London · New York ·
New Delhi · Bangkok · Singapore · Tokyo · Sydney

Oxidants and Antioxidants in Cutaneous Biology

Volume Editors *Jens Thiele*, Jena
Peter Elsner, Jena

36 figures, and 5 tables, 2001

Basel · Freiburg · Paris · London · New York ·
New Delhi · Bangkok · Singapore · Tokyo · Sydney

....
Current Problems in Dermatology

Library of Congress Cataloging-in-Publication Data

Oxidants and antioxidants in cutaneous biology / volume editors, Jens Thiele, Peter Elsner.
p.; cm. – (Current problems in dermatology; vol. 29)
Includes bibliographical references and index.
ISBN 3805571321 (hard cover : alk. paper)
1. Skin – Pathophysiology. 2. Free radicals (Chemistry) – Pathophysiology. 3. Antioxidants. 4. Active oxygen – Physiological effect. I. Thiele, Jens. II. Elsner, Peter, 1955– III. Series.
[DNLM: 1. Oxidants – adverse effects. 2. Skin Diseases – physiopathology. 3. Antioxidants – therapeutic use. 4. DNA Damage. 5. Oxidative Stress. 6. Signal Transduction. 7. Skin – radiation effects. 8. Ultraviolet Rays – adverse effects. WR 140 O973 2000]
RL96.O947 2000
616.5′07–dc21

00–048668

Bibliographic Indices. This publication is listed in bibliographic services, including Current Contents® and Index Medicus.

Drug Dosage. The authors and the publisher have exerted every effort to ensure that drug selection and dosage set forth in this text are in accord with current recommendations and practice at the time of publication. However, in view of ongoing research, changes in government regulations, and the constant flow of information relating to drug therapy and drug reactions, the reader is urged to check the package insert for each drug for any change in indications and dosage and for added warnings and precautions. This is particularly important when the recommended agent is a new and/or infrequently employed drug.

Printed in Switzerland on acid-free paper by Reinhardt Druck, Basel
ISBN 3–8055–7132–1

Dedication

This book is dedicated to Dr. *Lester Packer*, Professor of Physiology at the University of California, Berkeley, since 1961, and his wife *Anne*.

Together, and each in their own way, they have nurtured the careers of countless scientists, including four authors of this book.

Contents

Preface

Are free radicals and reactive oxygen species relevant for dermatopathology? Do antioxidants really protect against free-radical-mediated cutaneous disease and aging? In the past decade, a strongly increasing number of scientific publications on oxidative stresss and redox regulation in skin indicates the emerging importance of this field in experimental dermatology.

Furthermore, the increase in scientific evidence for protective antioxidative mechanisms in the skin has led to a considerably growing interest of the pharmaceutical and cosmetic industry in therapeutic antioxidant strategies. Likewise, such terms as 'free radicals', 'antioxidants' and 'oxidative stress' have experienced an almost inflationary use in the lay press and thus raised enormous interest in the public. Owing to this growing public, academic and corporate demand, this book is intended to provide an up-to-date overview of oxidants and antioxidants in cutaneous biology. This book compiles contributions from leading investigators on the detection of free radicals and antioxidants, their responses to environmental oxidative stressors, the role of oxidative DNA damage, UVB- and UVA-induced signal transduction, and antioxidant protection strategies. The chapters mainly focus on the outermost organ of the body, the skin. Thus, it represents a unique collection of important new facts and background information on oxidative-stress-related biochemistry, photobiology, molecular biology, pharmacology and cosmetology of the skin. The editors are indebted to all authors for the knowledge and effort they have invested in this project.

We sincerely hope that this book will provide valuable advice to our readers and thus will stimulate further the discovery of relevant redox-regulated pathways and the development of potent therapeutic antioxidant strategies in dermatology.

Jens Thiele, MD, Jena
Peter Elsner, MD, Jena

Thiele J, Elsner P (eds): Oxidants and Antioxidants in Cutaneous Biology.
Curr Probl Dermatol. Basel, Karger, 2001, vol 29, pp 1–17

Detection of Free Radicals in Skin: A Review of the Literature and New Developments

Jürgen Fuchs[a], *Thomas Herrling*[b], *Norbert Groth*[b]

[a] Department of Dermatology, Medical School, J.W. Goethe University, Frankfurt, and
[b] Center of Scientific Instruments, Laboratory for EPR Tomography, Berlin, Germany

It is generally believed that free radicals play an important role in the pathogenesis of several human diseases. When enough people believe in something, it becomes known or accepted. However, an increased generation of free radicals may be the consequence of tissue damage, an epiphenomenon or only of limited clinical significance. If free radicals are produced in a clinical condition, it has to be proved that these species are formed and are an obligatory intermediate for the pathology [1]. Of the several methods available to study free radicals in biological systems, electron paramagnetic resonance (EPR) spectroscopy is known to be the most important technique. Using the highly selective EPR technique, free radicals can be detected, characterized and quantified in biological systems. EPR spectroscopy is concerned with the resonant absorption of microwave radiation by paramagnetic samples in the presence of an applied magnetic field. Free radicals are paramagnetic species due to the unpaired electron in the outer orbit and have a magnetic moment. If an external magnetic field is applied to these molecules, their axes are directed either parallel (energetically more stable) to the external field or in the opposite direction (antiparallel). If electromagnetic waves which match the energy difference between the parallel and antiparallel electronic moments (microwaves) are applied to this system, a change in the orientation of these molecules will occur. The net absorption of the microwave energy under these resonance conditions is quantitated and the 2nd derivative is recorded as the EPR signal. Soon after the discovery of the EPR methodology by the Russian student Zavoiksy in 1945 [2], the method was applied for detection of free radicals in biological samples [3]. Since then, an uncountable number of reports on the development and application of EPR spectroscopy in biomedicine has appeared. One of the most comprehensive compilations of the EPR literature was given annually in the 'Specialist Periodical Reports by the Royal Society of Chemistry' [4]. The skin is a target organ of oxidative stress; because it is continuously exposed to high oxygen concentrations and solar radiation, it serves as a major portal of entry for many oxidizing environmental pollutants and occupational hazards, and contains several readily oxidizable molecules critical for structure and function. For these and other practical reasons outlined below, free radical biologists are becoming more and more interested in cutaneous EPR applications. The purpose of this review is to provide the reader with current information on the developments of free radical detection in skin by the EPR methodology and other applications of the EPR technique in cutaneous biology.

Direct Detection of Free Radicals

Direct evidence for free radical formation in human and animal skin following exposure to UV radiation has been obtained by low-temperature (e.g. –196 °C) EPR spectroscopy in vitro [5–7]. At the temperature of liquid nitrogen, the radical steady-state concentration is high enough to give a sufficient EPR signal intensity. However, these signals are very broad and usually provide no or only very limited information about the chemical identity of the free radical structure. At ambient temperature, only persistent free radicals such as melanin [8–14] or the ascorbyl radical [15–18] have been directly detected in skin and/or in skin appendages in vitro. The cutaneous metabolism of free-radical-generating compounds such as 4-hydroxyanisole and anthralin has been shown to generate persistent semiquinone radicals in rat skin [19] or anthrone/anthrone-dimer-derived radicals in pig or mouse skin in vitro [20, 21], respectively. Tocopheroxyl radicals were detected in vitamin-E-treated mouse skin in vitro after UV irradiation [22]. Some investigators have analyzed lyophilized or abraded skin samples (e.g. horn) for free radicals. The freeze-dry technique and mechanical processes such as cutting and grinding can produce paramagnetic artifacts in biological tissues [23, 24]. These techniques should be avoided, because they can lead to erroneous results in EPR experiments.

Indirect Detection of Free Radicals

In most cases, the low steady-state concentration of reactive free radicals in biological samples allows only indirect detection by the spin trapping method. Spin trapping is defined as that chemical reaction in which a radical adds to a molecule so that the group that was the radical (radical addend) stays with the molecule for future analysis. The molecule which captures the radical is called the spin trap. The additional product is called the spin adduct. Spin traps are usually nitrones or nitroso compounds, which are one electronic oxidation state above the nitroxides and which, upon reacting with free radicals, become converted to nitroxides. The trapped radical has a characteristic EPR spectrum that allows chemical identification of the highly reactive free radical. The interpretation of the EPR spectra and the subsequent identification of the free radical adduct is the most sophisticated and difficult component of the spin trapping process [1, 25, 26]. The stability of the radical adduct and its conversion into other paramagnetic or diamagnetic products can be a source of experimental confusion. The history of spin trapping began with C-phenyl-N-tert-butylnitrone (PBN), which is used for detection of carbon- and oxygen-centered free radicals [1, 26]. The capabilities of PBN were improved by synthesizing the pyridine-N-oxide analogue of PBN, C-(4-pyridinyl-N-oxide)-N-tert-butylnitrone (4-POBN). Dithiocarbamate iron (II) complexes are successfully used for spin trapping NO in biological systems [27, 28], while 2,2,6,6-tetramethyl-piperidine and 2,2,6,6-tetramethyl-4-piperidone are sensitive trapping agents for singlet oxygen [29, 30]. 5,5-Dimethyl-1-pyrroline-N-oxide (DMPO) is probably the most widely used spin trap in biological systems, scavenging carbon-, oxygen- and sulfur-centered free radicals [1, 27]. Table 1 shows some selected examples of spin trapping in keratinocytes, epidermis homogenate and skin biopsies. The free radical xenobiotic metabolism can be investigated by the spin trapping technique in animals in vivo, and some selected examples of in vivo spin trapping are shown in table 2. To our knowledge however, until now no in vivo studies in skin have yet been published.

Nitroxide-Based Electron Paramagnetic Resonance

Because of their physicochemical versatility, nitroxides can be used as imaging agents for a number of different purposes including the investigation of the cellular redox status, structural and dynamic properties of biological membranes, oximetry and pH measurements.

Redox Measurements

Nitroxides can be used for the study of redox metabolism [45–47]. Nitroxide free radicals accept electrons from a variety of sources such as low-molecular-weight antioxidants, certain enzymes such as the cytochrome P_{450} system,

Table 1. In vitro spin trapping in keratinocytes and epidermis homogenate

Free radical or reactive species detected	Sample	Spin trap	Reference
Alkyl hydroxyl	rat epidermis homogenate	DMPO	7
Lipid alkyl Lipid alkoxyl	mouse skin biospy human skin biopsy	DMPO 4-POBN	17
Alkyl Alkoxyl	mouse skin biopsy	DMPO	31
Singlet oxygen	human bronchial epithelial cells	2-(9,10-dimethoxyanthracenyl)-tert-butylhydroxylamine	32
Glutathione thiyl	keratinocytes	DMPO	33
Alkyl Alkoxyl	mouse keratinocytes	DMPO	34
Hydroxyl	murine skin fibroblasts	DMPO	35
Methyl	human keratinocytes	4-POBN	36
Methyl	human squamous carcinoma keratinocytes	3,5-dibromonitrosobenzene sulfonic acid	37
Hydroxyl	guinea pig epidermis homogenate	DMPO	38
Lipid alkyl hydroxyl	rat epidermis homogenate	DMPO	39
Carbon-centered	mouse skin biopsy	4-POBN	40

the mitochondrial respiratory chain and transition metal ions. They also react with reactive oxidants, such as the superoxide anion radical [48, 49], and other free radical species [50]. Thus, the biokinetics of nitroxides is sensitive to the reducing as well as to the oxidizing activity of their ultimate surrounding. The distribution of the nitroxides as well as the heterogenous cellular and subcellular distribution of different reducing and oxidizing agents must be taken into account, when analyzing nitroxide biokinetics. Nitroxide-based in vitro EPR studies for the measurement of redox components in isolated keratinocytes, skin homogenates and intact skin samples have been published [51–55]. As outlined above, the signal decay rate of nitroxides is enhanced by oxidative stress, and this enhanced decay can be suppressed by administration of antioxidants. Due to their redox properties 3-carbamoyl-2,2,5,5-tetramethyl-pyrrolidine-N-yloxyl (CTPO) is preferentially used to study the effect of reactive oxidants, while 2,2,6,6-tetramethyl-piperidine-N-oxyl (TEMPO) is employed

Table 2. In vivo spin trapping

Reactive species	Sample	Spin trap	EPR technique	Reference
Sulfur trioxide anion radical	whole-mouse sequential intravenous injections of sodium sulfite and sodium dichromate	DMPO	1 GHz	41
Oxygen-centered radicals	blood effluate from ischemic skin flap	DMPO	1 GHz	42
Nitric oxide	subcutaneous compartment of mice, injected with isosorbite dinitrate	N-(dithiocarboxy) sarcosine	1 GHz	43
Hydroxyl radical	subcutaneous mouse tumor, ionizing irradiation	DMPO	1 GHz	44

for measurement of tissue antioxidant activity. Nitroxides have been used as indicators of the cellular redox status for in vivo measurement of antioxidant status as well as oxidative stress as shown in table 3.

Oximetry

EPR studies can give information on the tissue oxygen tension by virtue of a physical interaction of molecular oxygen, which is paramagnetic, and a spin probe thereby modifying the spectral characteristic of the spin probe. The extent of spectral broadening can usually be directly correlated to oxygen concentration by appropriate calibrations. Spin probes that have been used for tissue oximetry include nitroxides, lithium phthalocyanine, India ink, fusinite, synthetic chars and coals. This method provides a sensitivity, accuracy and range to measure physiologically and pathologically pertinent oxygen tensions in vivo. For illustration, a solid-state paramagnetic probe such as lithium phthalocyanine [66] or charcoal [67] has been used in localized tissue oximetry, and nitroxide spin probes were used for measurement of tissue oxygen tension in ischemic tissue [68]. Low-frequency EPR (250 MHz) was successfully utilized to measure oxygen tension in tumor tissues of living mice using a perdeuterated nitroxide spin probe [69]. To date there are only a few reports of EPR oximetry in skin [70, 71]. It seems likely that this will change in the near future, because skin oxygenation greatly influences the effectiveness of many anticancer therapies such as chemotherapy, radiation therapy, hyperthermia and photodynamic therapy. The effectiveness of these therapies can be manipulated by modulating

Table 3. Nitroxides as indicators of the cellular redox status for in vivo measurement of antioxidant status as well as oxidative stress

Antioxidant/oxidant modulation of tissue redox status	Nitroxide	Sample	EPR technique	Reference
Antioxidant supplementation	TEMPO	mouse lung	1 GHz	56
Antioxidant supplementation	TEMPO	rat	1 GHz	57
Antioxidant status UV irradiation	TEMPO	human skin	3 GHz	58
Streptozotocin-induced diabetes	CTPO	rat	300 MHz	59
Silica-induced lung injury	TEMPO	rat lung	1 GHz	60
Hyperoxia	CTPO	mouse	1 GHz, loop gap resonator	61
Ionizing radiation	CTPO	mouse	1 GHz, loop gap resonator	62
Iron overload	CTPO	mouse	1 GHz, loop gap resonator	63
Ionizing radiation	CTPO	mouse	1 GHz	64
Carbon tetrachloride	CTPO	mouse	1 GHz	65

skin oxygen tension. It may be possible to utilize EPR-based oximetry as an in situ predictor of the energy/dose required to elicit a biological response in skin.

Membrane Structure

Spin labeling is an EPR technique used to monitor biophysical properties of biological membranes. For example, this is achieved by introducing a nitroxide-labeled fatty acid into the membrane system. The nitroxide group is sensitive to its biophysical surrounding. Thereby fluidity and polarity of membranes can be analyzed in complex biological samples. Since the lamellarly arranged lipid bilayers of the stratum corneum control the diffusion and penetration of chemical substances into and through the skin, EPR-based measurement of stratum corneum lipid microviscosity and polarity provides information on the barrier function of the epidermis. An increase in fluidity of skin lipid bilayers suggests a decrease in the skin barrier function. The pH value of the stratum corneum is an important regulating factor for the stratum corneum homeostasis, and it is assumed that the pH is among the factors that regulate the integrity of the skin barrier function. The use of pH-sensitive nitroxides, in conjunction with EPR, offers a unique opportunity for noninvasive assessment of pH values in vivo [72]; however, no such studies have yet been performed in skin. The fluidity of animal and human skin-derived stratum corneum lipids was analyzed in vitro by the EPR technique employing nitroxide-labeled 5(12,16)-doxylstearic acid [3–77] and perdeuterated di-tert-butylnitroxide [78]. It was suggested that EPR may provide a facile and robust method to define the subclinical irritancy potential of chemicals [75].

Electron Paramagnetic Resonance Imaging

The spatial distribution of free radicals within a biological sample can be analyzed by utilizing magnetic field gradients in a manner similar to that of NMR imaging [79, 80]. EPR imaging (EPRI) and NMR imaging are based on similar principles. However, the superiority of NMR imaging is due to its excellent sensitivity. In vivo EPRI is presently restricted to artificial free radicals introduced into a biological sample with a sensitivity of tissue free radical concentration less than 0.1 m*M* compared to more than 100 *M* of endogenous proton concentrations available for NMR. EPRI can be performed in the spatial range to obtain one-, two- or three-dimensional images of free radical distribution in samples. The imaging technique that also includes a spectral dimension is termed spectral-spatial imaging. Spectral-spatial imaging can also be performed in one, two or three spatial dimensions. The spectral-spatial image contains more information, but it is technically more difficult to obtain. We have used the EPRI technique with modulated field gradients to obtain spatial resolution of paramagnetic centers in different tissue planes. This approach allows the measurement of an EPR spectrum in a selected volume part [81]. EPRI at 9 GHz was used in vitro for measuring biokinetics and spatial or spectral-spatial distribution of nitroxides in mouse or pig skin [82–87]. Penetration of nitroxide-labeled drugs such as retinoic acid and dihydrolipoate was studied by this technique in mouse skin biopsies [88, 89]. Although there are some limitations of the spin labeling method, drug penetration as well as drug-membrane and drug-enzyme interactions in skin can be investigated by labeling pharmacologically active compounds.

The working frequency, the radical concentration and the magnetic field gradient determine the spatial resolution of EPRI. The application of a single line paramagnetic label yielded an image resolution better than 100 μm at 1 GHz on samples of up to 20 mm in size [90]. Berliner et al. [91] measured for the first time in vivo an EPR image of a murine tumor (Cloudman S-91 melanoma in the tail of a DBA-2J mouse) using nitroxide injected into the tail vein. They obtained a cross-sectional image perpendicular to the tail

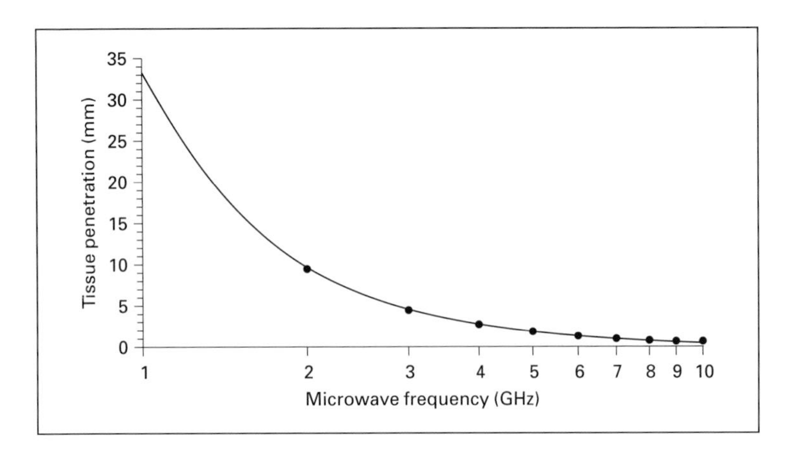

Fig. 1. Penetration depth of microwaves into biological tissues as a function of different frequencies.

axis, which clearly distinguished features to the submillimeter resolution level [91]. The application of the EPRI technique in vivo to obtain high-quality images of biological samples is presently limited by several factors including gradient design and accuracy, sensitivity as well as speed of acquisition. A further intrinsic problem of the EPRI technique is the short relaxation time of the free electron. The line width associated with EPR signals is 3 orders of magnitude larger compared to that of NMR signals. Therefore EPRI requires 100–1,000 times more powerful gradients, and this significantly decreases the sensitivity, requiring a tissue spin probe concentration of at least 0.1 m*M*.

In vivo Electron Paramagnetic Resonance

In vivo applications of EPR are hampered by the limited tissue penetration depths of microwaves and the high nonresonant dielectric loss of the exciting frequency. Due to the high water content of biological samples, the penetration depth of the most commonly applied microwave frequency of 9 GHz is less than 1 mm [92]. Figure 1 shows the penetration depth of microwaves into biological tissues as a function of different frequencies, indicating that low microwave frequencies have a better tissue penetration than high-frequency radiation. The nonresonant dielectric absorption of microwaves in biological samples is a function of the frequency. 9-GHz EPR measurements on lossy samples are limited to a volume of few microliters, 1- to 3-GHz measurements are possible on volumes of a few milliliters and radiofrequency resonators (200–300 MHz) are used for samples of up to 200 ml. Depending on the special characteristics of the EPR instrument and sample cavity used, skin biopsies 4–6 mm in diameter are the maximum size that allows for adequate tuning using a conventional H_10_2 cavity at 9 GHz frequency [93]. The strength of the EPR signal is proportional to the product of the quality factor and the filling factor of the cavity. Resonator cavity design is critical to achieve maximum sensitivity and must be adapted to accommodate the sample with the highest possible filling factor × quality factor product. The application of loop gap and reentrant resonators provides an excellent approach to achieve this. Surface coil type loop gap resonators were successfully used for imaging the subcutaneous compartment of mice [94] or subcutaneously localized tumors in mice [95] in vivo. Due to their better tissue-penetrating properties, microwave frequencies lower than 3 GHz permit in vivo EPR studies in whole animals [96, 97]. Whole-body EPRI in small animals measuring nitroxide radical elimination was successfully performed at very low microwave frequency (200–700 MHz) [98, 99] and at low frequency (1 GHz) [100, 101]. However, a severe disadvantage of reducing the microwave frequency is a significant loss in sensitivity. The sensitivity of the measurement is directly related to the square of the operating frequency. For practical reasons, at 9 GHz frequency, the detectability limit of EPR spectroscopy using a conventional H_10_2 cavity is in the range of 10^{-6}–10^{-7} *M*, although lower radical steady-state concentrations can be detected in biological samples by accumulation techniques, modifications of the cavity or more sensitive detection methods. A further loss in sensitivity (100- to 1,000-fold) occurs by application of the imaging technique in comparison to spectroscopy, depending on the image reconstruction method. This explains that the usually applied nitroxide concentrations in small animal in vivo EPRI experiments are in the range of 0.1–10 m*M* and higher. Technical innovations of in vivo EPR have been the integration and combination of different magnetic resonance techniques, such as the development of proton electron double resonance imaging (PEDRI) and field-cycled dynamic nuclear polarization (FC-DNP). At low microwave frequency PEDRI and FC-DNP analyze the electron spins indirectly via their effect on a proton NMR signal [102–104]. Both techniques require administration of artificial spins (e.g. nitroxide radicals) and almost all the machines used for PEDRI and FC-DNP have been constructed for small animals. As already outlined above, the most restricting factors for clinical applications are the limited tissue penetration of the microwaves and the weak EPR effect of low-frequency microwave radiation causing a significant loss in sensitivity. An approach to improve the sensitivity is the development of new detection techniques for the EPR resonance phenomenon such as the design of instrumentation with

Table 4. In vivo skin EPR spectroscopy and imaging

Nitroxide	Organ	Species	EPR technique	Reference
TEMPO	skin	human	3-GHz spectroscopy	58
CTPO	melanoma	mouse	1.5-GHz imaging	83
Polynitroxyl albumin + 4-hydroxy-2,2,6,6-tetramethyl-piperidine-N-oxyl	subcutaneous tissue	mouse	1-GHz spectroscopy and imaging	94
Perdeuterated N^{15} TEMPO and CTPO	subcutaneous tissue	mouse	1-GHz spectroscopy and imaging	95
Perdeuterated N^{15} CTPO	skin	mouse	3-GHz spectroscopy	107
TEMPO	skin	human	3-GHz spectroscopy	108, 109
CTPO	skin	mouse	9-GHz spectroscopy	110
Di-tert-butylnitroxide	tail vasculature	mouse	9-GHz spectroscopy	111
CTPO	subcutaneous tissue	rat	700-MHz spectroscopy, flexible surface coil type resonator (stethoscope-like application)	112
Anthralin-derived free radicals	skin	mouse	1-GHz spectroscopy	113
Cr(V)	skin	rat	1-GHz spectroscopy	114

longitudinally detected EPR [105]. Other techniques comprise photoacoustic detection of magnetic resonance, detecting magnetic flux changes due to resonance by a superconductive split ring, Raman-heterodyne EPR, electrical detection of EPR signals (ED-EPR and STM-EPR) and fluorescence-detected magnetic resonance [106]. ED-EPR, STM-EPR and fluorescence-detected magnetic resonance have been used for the detection of very strong paramagnetic centers such as in ferritin or silicium, but it is questionable whether these techniques can be applied to biological samples. However, further developments of these detection techniques may lead to a significant improvement in the sensitivity, which is presently a limiting factor for many EPR applications in vivo.

Because of its location, the skin is fully accessible to relatively higher-frequency EPR (eg. 3–9 GHz), in contrast to many other sites where the depth of sensitivity can be limiting. Presently, the skin is the only human organ which can be measured in vivo with sufficient sensitivity. Table 4 shows some selected examples of in vivo EPR spectroscopy and imaging in animal and human skin. Two different experimental arrangements have been used for in vivo measurements on skin, and one technique is presently being developed for localized skin EPR measurements of the human body. For the microwave system, either a 9-GHz bridge with a cavity or a 3-GHz bridge with a surface coil are used. The 9-GHz bridge system is based on the concept of Furusawa and Ikeya [115], who used microwave cavities with a small hole from which the microwave field leaks out to a small cross-sectional area of the object. This method is also applicable for localized spectroscopy on skin. For the magnet system, a normal electromagnet with a gap of 100 mm is wide enough to accommodate human limbs between the pole faces. Figure 1 shows that the microwave penetration depths lie between 0.5–1.0 mm (9 GHz) and about 5 mm (3 GHz), respectively. This means that 9 GHz is restricted to the upper layer of the skin (i.e. the human epidermis and upper dermis, or total mouse skin). For deeper layers (i.e. the full human dermis or subcutis) penetration depths of 5 mm (3 GHz) or 3.5 cm (1 GHz) are more suited, respectively. The probe head is a 90-degree bent surface coil (8 mm diameter) with an electronically matched system. A quartz plate is mounted on one side of the surface coil, which defines a plane-parallel measuring area on the skin. Matching is accomplished by placing a piezoelectric element at a distance of $^1/_4$ outside the loop. Two 100-kHz modulation coils near the surface coil generate a modulation field Bm in the skin layer. The rapid scan coils are mounted on the surface of the magnet pole plates. For human measurements, the probe head is mounted on the forearm placed in the magnet. It is important to point out that this apparatus is restricted to human limbs and does not allow measurements on other parts of the body such as the breast or spine which would require a more flexible probe head accessible to all parts of the body. Since magnetic field requirements for EPR, such as strength, homogeneity and stability of the magnetic field, are much lower than for NMR, this allows the use of a compact and flexible probe head of the size used in sonography, easily permitting skin measurements on all parts of the human body without requiring a whole-body magnet. Such a flexible compact probe head is presently under construction in our laboratory. The prototype of this newly designed probe head consists of two rectangular pieces of Neo Delta Magnet material (Nd Fe B), connected by an iron backbone. Two small magnet pieces near the microwave loop increase the homogeneity of the magnetic field in the upper skin layers. A field sweep of 6 mT is generated by two coils wound on the backbone of the magnet system. The microwave shield above the microwave loop prevents interactions between the loop and the magnet. This flexible and compact probe head will allow nitroxide and spin-trap-based EPR spectroscopy in vivo on all parts of the human skin with high specificity.

Outlook

The growing interest in the role of free radicals in the pathogenesis of human diseases has led to an increased need for techniques to measure free radicals in the clinical situation [116]. EPR spectroscopy is a highly selective assay for detecting free radicals and is the most important technique for characterization and quantification of these species in biological systems. EPRI can be performed in the spatial range to obtain images of free radical distribution in samples, while spectral-spatial imaging includes a spectral dimension. The success of EPR-based studies relies heavily on the use of spin traps or nitroxides, because the steady-state concentration of most endogenous free radicals is orders of magnitude below the detectability limit of EPR. Although in the last few years EPR spectroscopy and EPRI techniques have been considerably developed to give useful biochemical and biophysical information in vivo, these methodologies need to be improved for reliable application under clinical conditions, as several intrinsic technical problems must still be resolved. Human in vivo EPR is hampered by a significant loss in sensitivity when using more deeply penetrating low-frequency microwaves and by practical problems of the sample and magnet/resonator size. It was guessed that at least one or two orders of magnitude of higher sensitivity are necessary to detect reactive free radicals directly in tissue [106]. However, this is presumably an optimistic calculation. Improved detection techniques and the development of high-quality surface coil resonators could help to resolve some of the difficulties. The use of in vivo EPR to study metabolic processes in skin appears to be an attractive and effective approach, because of the importance of this organ and its acccessibility. The skin is presently the only human organ which can be measured in vivo with sufficient sensitivity at 9 and 3 GHz frequency. The more widespread application of localized EPR spectroscopy in human skin in vivo will significantly contribute to improve our understanding of cutaneous free radical processes and redox biochemistry. Furthermore, the EPR methodology is a useful tool for the noninvasive in vivo measurement of skin barrier function, drug/skin interaction and cutaneous oxygen tension.

References

1 Janzen EG: Spin trapping; in Ohya-Nishiguchi H, Packer L (eds): Molecular and Cell Biology Updates: Bioradicals Detected by ESR Spectroscopy. Basel, Birkhäuser, 1995, pp 113–142.

2 Zavoisky E: Paramagnetic relaxation of liquid solutions for perpendicular fields. J Physics (USSR) 1945;9:211–216.

3 Commoner B, Townsend J, Pake GW: Free radicals in biological materials. Nature 1954;174:689–691.

4 Symons MCR (ed): Electron Spin Resonance: A Specialist Periodical Report. London, Royal Society of Chemistry, Burlington House, 1989, vol 13.

5 Norins AL: Free radical formation in the skin following exposure to ultraviolet light. J Invest Dermatol 1962;39:445–448.
6 Pathak MA, Stratton K: Free radicals in human skin before and after exposure to light. Arch Biochem Biophys 1968;123:468–476.
7 Nishi J, Ogura R, Sugiyama M, Kohno M: Involvement of active oxygen in lipid peroxide radical reaction of epidermis following ultraviolet light exposure. J Invest Dermatol 1991;97:115–119.
8 Blois MS, Zahlan AB, Maling JE: Electron spin resonance studies on melanin. Biophys J 1964;4: 471–490.
9 Sealy RC, Felix CC, Hyde JS, Swartz HM: Structure and reactivity of melanins: Influence of free radicals and metal ions. Free Radic Biol 1980;4:209–259.
10 Arnaud R, Perbet G, Deflandre A, Lang G: Electron spin resonance of melanin from hair: Effects of temperature, pH and light irradiation. Photochem Photobiol 1983;38:161–168.
11 Sarna T, Korytowski W, Sealy RC: Nitroxides as redox probes of melanins: Dark induced and photoinduced changes in redox equilibria. Arch Biochem Biophys 1985;239:226–233.
12 Vsevolodov EB, Ito S, Wakamatsu K, Kuchina II, Latypov IF: Comparative analysis of hair melanins by chemical and electron spin resonance methods. Pigment Cell Res 1991;4:30–34.
13 Slominski A, Paus R, Plonka P, Chakraborty A: Melanogenesis during the anagen-catagen-telogen transformation of the murine hair cycle. J Invest Dermatol 1994;102:862–869.
14 Collins B, Poehler TO, Bryden WA: EPR persistence measurements of UV-induced melanin free radicals in whole skin. Photochem Photobiol 1995;62:557–560.
15 Buettner GR, Motten AG, Hall RD, Chignell CF: ESR detection of endogenous ascorbate free radical in mouse skin: Enhancement of radical production during UV irradiation following topical application of chlorpromazine. Photochem Photobiol 1987;46:161–164.
16 Jurkiewicz BA, Buettner GR: Ultraviolet light induced free radical formation in skin: An electron paramagnetic resonance study. Photochem Photobiol 1994;59:1–4.
17 Jurkiewicz BA, Buettner GR: EPR detection of free radicals in UV irradiated skin: Mouse versus human. Photochem Photobiol 1996;64:918–935.
18 Haywood RM, Wardman P, Gault DT, Linge C: Ruby laser irradiation (694 nm) of human skin biopsies: Assessment by electron spin resonance spectroscopy of free radical production and oxidative stress during laser depilation. Photochem Photobiol 1999;70:348–352.
19 Riley PA: Mechanism of pigment cell toxicity produced by hydroxyanisole. J Pathol 1970;101: 163–169.
20 Shroot B, Brown C: Free radicals in skin exposed to dithranol and its derivatives. Drug Res 1986; 36:1253–1255.
21 Fuchs J, Packer L: Investigations on anthralin free radicals in model systems and in skin of hairless mice. J Invest Dermatol 1989;92:677–682.
22 Fuchs J: Elektronen paramagnetische Resonanz-Spektroskopie und Tomographie in der experimentellen Dermatologie; Habilitationsschrift für das Fach Dermatologie und Venerologie; Fachbereich Humanmedizin der Johann-Wolfgang-Goethe-Universität, Frankfurt/M, 1992.
23 Heckly RJ: Free radicals in dry biological systems. Free Radic Biol 1980;4:135–158.
24 Symons MCR: Formation of radicals by mechanical processes. Free Radic Res 1988;5:131–139.
25 Buettner GR: Spin trapping: ESR parameters of spin adducts. Free Radic Biol Med 1987;3:259–303.
26 Brackett DJ, Wallis G, Wilson MF, McCay PB: Spin trapping and electron paramagnetic resonance spectroscopy; in Armstrong D (ed): Methods in Molecular Biology: Free Radical and Antioxidant Protocols. Totowa, Humana Press, 1997, vol 108, pp 15–25.
27 Tsuchiya K, Jiang JJ, Yoshizumi M, Tamaki T, Houchi H, Minakuchi K, Fukuzawa K, Mason RP: Nitric oxide-forming reactions of the water soluble nitric oxide spin trapping agent MGP. Free Radic Biol Med 1999;27:347–355.
28 Nakagawa H, Ikota N, Ozawa T, Masumizu T, Kohno M: Spin trapping for nitric oxide produced in LPS-treated mouse using various new dithiocarbamate iron complexes having substituted proline and serine moiety. Biochem Mol Biol 1998;45:1129–1138.
29 Ando T, Yoshikawa T, Tanigawa T, Kohno M, Yoshida N, Kondo M: Quantification of singlet oxygen from hematoporphyrin derivative by electron spin resonance. Life Sci 1997;61:1953–1959.

30 Konaka R, Kasahara E, Dunlap WC, Yamamoto Y, Chien KC, Inoue M: Irradiation of titanium dioxide generates both singlet oxygen and superoxide anion. Free Radic Biol Med 1999;27:294–300.

31 Timmins GS, Davies MJ: Free radical formation in murine skin treated with tumour promoting organic peroxides. Carcinogenesis 1993;14:1499–1503.

32 Nye AC, Rosen GM, Gabrielson EW, Keana JFW, Prabhu VS: Diffusion of singlet oxygen into bronchial epithelial cells. Biochim Biophys Acta 1987;928:1–7.

33 Schreiber J, Foureman GL, Hughes MF, Mason RP, Eling TE: Detection of glutathione thiyl free radical catalyzed by prostaglandin H synthase present in keratinocytes. J Biol Chem 1989;264: 7936–7943.

34 Taffe BG, Takahashi N, Kensler TW, Mason RP: Generation of free radicals from organic hydroperoxide tumor promoters in isolated mouse keratinocytes. J Biol Chem 1987;262:12143–12149.

35 Masaki H, Sakurai H: Increased generation of hydrogen peroxide possibly from mitochondrial respiratory chain after UVB irradiation of murine fibroblasts. J Dermatol Sci 1997;14:207–216.

36 Iannone A, Marconi A, Zambruno G, Gianetti A, Vannini V, Tomasi A: Free radical production during metabolism of organic hydroperoxides by normal human keratinocytes. J Invest Dermatol 1993;101:59–63.

37 Athar M, Mukhtar H, Bickers DR, Khan IU, Kalyanaraman B: Evidence for the metabolism of tumor promoter organic hydroperoxides into free radicals by human carcinoma skin keratinocytes: An ESR-spin trapping study. Carcinogenesis 1989;10:1499–1503.

38 Taira J, Mimura K, Yoneya T, Hagi A, Murakami A, Makino K: Hydroxyl radical formation by UV-irradiated epidermal cells. J Biochem (Tokyo) 1992;111:693–695.

39 Ogura R, Sugiyama M, Nishi J, Haramaki N: Mechanism of lipid radical formation following exposure of epidermal homogenate to ultraviolet light. J Invest Dermatol 1991;97:1044–1047.

40 Jurkiewicz BA, Bissett DL, Buettner GR: Effect of topically applied tocopherol on ultraviolet radiation mediated free radical damage in skin. J Invest Dermatol 1995;104:484–488.

41 Jiang J, Liu KJ, Shi X, Swartz HM: Detection of short-lived free radicals by low-frequency electron paramagnetic resonance spin trapping in whole living animals. Arch Biochem Biophys 1995;319: 570–575.

42 Miller CW, Chen G, Janzen EG: Detection of free radicals in reperfused dog skin flaps using electron paramagnetic resonance spectroscopy: A pilot study. Microsurgery 1999;19:171–175.

43 Fujii S, Suzuki Y, Yoshimura T, Kamada H: In vivo three-dimensional EPR imaging of nitric oxide production from isosorbide dinitrate in mice. Am J Physiol 1998;274:G857–G862.

44 Halpern HJ, Yu C, Barth E, Peric M, Rosen GM: In situ detection by spin trapping of hydroxyl radical markers produced from ionizing radiation in the tumor of living mice. Proc Natl Acad Sci USA 1995;92:796–800.

45 Swartz HM: Use of nitroxides to measure redox metabolism in cells and tissues. J Chem Soc Farady Trans I 1987;83:191–202.

46 Sotgiu A, Colacicchi S, Placidi G, Alecci M: Water soluble free radicals as biologically responsive agents in electron paramagnetic resonance imaging. Cell Mol Biol 1997;43:813–823.

47 Sutcliffe LH: The design of spin probes for electron magnetic resonance spectroscopy and imaging. Phys Med Biol 1998;43:1987–1993.

48 Voest EE, van Faassen E, Marx JJM: An electron paramagnetic resonance study of the antioxidant properties of the nitroxide free radical TEMPO. Free Radic Biol Med 1993;15:589–595.

49 Zhang R, Goldstein S, Samuni A: Kinetics of superoxide induced exchange among nitroxide antioxidants and their oxidized and reduced forms. Free Radic Biol Med 1999;26:1245–1252.

50 Asmus KD, Nigam S: Kinetics of nitroxyl radical reactions: A pulse-radiolysis conductivity study. Int J Radiat Biol 1976;29:211–219.

51 Fuchs J, Mehlhorn RJ, Packer L: Free radical reduction mechanisms in mouse epidermis and skin homogenates. J Invest Dermatol 1989;93:633–640.

52 Fuchs J, Huflejt ME, Rothfuss LM, Wilson DB, Gerardo C, Packer L: Impairment of enzymic and nonenzymic antioxidants in skin by photooxidative stress. J Invest Dermatol 1989;93:769–773.

53 Fuchs J, Mehlhorn RJ, Packer L: Assay of free radical reductase activity in biological tissue such as skin by ESR spectroscopy. Methods Enzymol 1990;186:670–674.

54 Fuchs J, Freisleben HJ, Podda M, Zimmer G, Milbradt R, Packer L: Nitroxide radical biostability in skin. Free Radic Biol Med 1993;15:415–423.
55 Fuchs J, Groth N, Herrling T, Zimmer G: Electron paramagnetic resonance studies on nitroxide radical 2,2,5,5-tetramethyl-4-piperidine-1-oxyl (Tempo) redox reactions in human skin. Free Radic Biol Med 1997;22:967–976.
56 Takeshita K, Hamada A, Utsumi H: Mechanisms related to reduction of radical in mouse lung using an L-band ESR spectrometer. Free Radic Biol Med 1999;26:951–960.
57 Matsumoto S, Mori N, Tsuchihashi N, Ogata T, Lin Y, Yohoyama H, Shin-Ichi I: Enhancement of nitroxide reducing activity in rats after chronic administration of vitamin E, vitamin C, and idebenone examined by an in vivo electron spin resonance technique. Magn Reson Med 1998;40: 330–333.
58 Herrling T, Zastrow L, Groth N, Fuchs J, Stanzl K: Detection and influencing of the antioxidative potential (AOP) of human skin. Seifen Öle Fette Wachse J 1996;122:472–476.
59 Sano T, Umeda F, Hashimoto T, Nawata N, Utsumi H: Oxidative stress measurement by in vivo electron spin resonance spectroscopy in rats with streptozotocin induced diabetes. Diabetologica 1998;41:1355–1360.
60 Vallyathan V, Leonhard S, Kuppusamy P, Pack D, Chzhan M, Sanders SP, Zweir JL: Oxidative stress in silicosis: Evidence for the enhanced clearance of free radicals from whole lungs. Mol Cell Biochem 1997;168:125–132.
61 Miura Y, Hamada A, Utsumi H: In vivo studies of antioxidant activity on free radical reaction in living mice under oxidative stress. Free Radic Res 1995;22:209–214.
62 Miura Y, Anzai K, Urano S, Ozawa T: In vivo electron paramagnetic resonance studies on oxidative stress caused by X-irradiation in whole mice. Free Radic Biol Med 1997;23:533–540.
63 Phumala N, Ide T, Utsumi H: Noninvasive evaluation of in vivo free radical reactions catalyzed by iron using in vivo ESR spectroscopy. Free Radic Biol Med 1999;26:1209–1217.
64 Utsumi H, Kawabe H, Masuda S, Takeshita K, Miura Y, Ozawa T, Hashimoto T, Ikehira H, Ando K, Yukawa O, Hamada A: In vivo ESR studies on radical reaction in whole mice – Effects of radiation exposure. Free Radic Res 1992;16:1–5.
65 Utsumi H, Ichikawa K, Takeshita K: In vivo ESR measurements of free radical reactions in living mice. Toxicol Lett 1995;82–83:561–565.
66 Liu KJ, Gast P, Moussavi M, Norby SW, Vahidi N, Walczak T, Wu M, Swartz HM: Lithium phtalocyanine: A probe for electron paramagnetic resonance oximetry in viable biological systems. Proc Natl Acad Sci USA 1993;90:5438–5442.
67 He G, Shankar RA, Chzhan M, Samoulion A, Kuppusamy P, Zweier JL: Noninvasive measurement of anatomic structure and intraluminal oxygenation in the gastrointestinal tract of living mice with spatial and spectral EPR imaging. Proc Natl Acad Sci USA 1999;96:4586–4591.
68 Zweier JL, Chzhan M, Ewert U, Schneider G, Kuppusamy P: Development of a highly sensitive probe for measuring oxygen in biological tissue. J Magn Res B 1994;105:52–57.
69 Halpern JH, Yu C, Peric M, Barth ED, Karczmar GS, River JN, Grdina DJ, Teicher BA: Measurement of differences in pO_2 in response to perfluorocarbon/carbogen in FSa and NFSa murine fibrosarcomas with low frequency electron paramagnetic resonance oximetry. Radiat Res 1996;145: 610–618.
70 Hatcher ME, Plachy WZ: Dioxygen diffusion in the stratum corneum: An EPR spin label study. Biochim Biophys Acta 1993;1149:73–78.
71 Swartz HM, Clarkson RB: The measurement of oxygen in vivo using EPT techniques. Phys Med Biol 1998;43:1957–1975.
72 Sotgiu A, Mäder K, Placidi G, Colacicchi S, Ursini CL, Alecci M: pH sensitive imaging by low frequency EPR: A model study for biological application. Phys Med Biol 1998;43:1921–1930.
73 Ogiso T, Hirota T, Iwaki M, Hino T, Tanino T: Effect of temperature on percutaneous absorption of terodiline, and relationship between penetration and fluidity of the stratum corneum lipids. Int J Pharm 1998;176:63–72.
74 Kitagawa S, Hosokai A, Kaseda Y, Yamamoto N, Kaneko Y, Matsuoka E: Permeability of benzoic acid derivatives in excised guinea pig dorsal skin and effects of L-menthol. Int J Pharm 1998;161: 115–122.

75 Kawasaki Y, Quan D, Sakamoto K, Maibach HI: Electron resonance studies on the influence of anionic surfactants on human skin. Dermatology 1997;194:238–242.
76 Gay CL, Murphy TM, Hadgraft J, Kellaway IW, Evans JC, Rowlands CC: An ESR study of skin penetration enhancers. Int J Pharm 1989;49:39–46.
77 Alonso A, Meirelles NC, Yushmanov VE, Tabak M: Water increases the fluidity of intercellular membranes of stratum corneum: Correlation with water permeability, elastic, and electrical resistance properties. J Invest Dermatol 1996;106:1058–1063.
78 Rehfeld SJ, Plachy WZ, Hou SYE, Elias PM: Localization of lipid microdomains and thermal phenomena in murine stratum corneum and isolated membrane complexes: An electron spin resonance study. J Invest Dermatol 1990;95:217–223.
79 Berliner LJ: The development and future of ESR imaging and related techniques. Phys Med 1989; 5:63–75.
80 Swartz HM, Walczak T: In-vivo EPR: Prospects for the 90s. Phys Med 1993;9:41–50.
81 Herrling T, Klimes N, Karthe W, Ewert U, Ewert B: EPR zeugmatography using modulated magnetic field gradients. J Magn Reson 1982;49:203–211.
82 Kristel J, Pecar S, Korbar-Smid J, Demsar F, Schara S: Drug diffusion: A field gradient electron paramagnetic resonance study. Drug Dev Ind Pharm 1989;15:1423–1440.
83 Gabrijelcic V, Sentjurc M, Kristl J: Evaluation of liposomes as drug carriers into the skin by one dimensional EPR imaging. Int J Pharm 1990;62:75–79.
84 Fuchs J, Milbradt R, Groth N, Herrling N, Zimmer G, Packer L: One- and two-dimensional EPR (electron paramagnetic resonance) imaging in skin. Free Radic Res 1991;15:245–253.
85 Fuchs J, Milbradt R, Groth N, Herrling T, Zimmer G, Packer L: Electron paramagnetic resonance (EPR) imaging in skin: Biophysical and biochemical microscopy. J Invest Dermatol 1992;98:713–719.
86 Herrling T, Groth N, Fuchs J: Biochemical EPR imaging of skin. Appl Magn Reson 1996;11: 471–486.
87 Sentjurc M, Vrhovnik K, Kristl J: Liposomes as a topical delivery system: The role of size on transport studied by the EPR imaging method. J Controlled Release 1999;59:87–97.
88 Michel C, Groth N, Herrling T, Rudolph P, Fuchs J, Kreuter J, Freisleben HJ: Penetration of spin-labeled retinoic acid from liposomal preparations into the skin of SKH1 hairless mice: Measurement by EPR tomography. Int J Pharm 1993;98:131–139.
89 Freisleben HJ, Groth N, Fuchs J, Rudolph P, Zimmer G, Herrling T: Penetration of spin labeled dihydrolipoate into the skin of hairless mice. Drug Res 1994;44:1047–1050.
90 Kuppusamy P, Wang P, Chzhan M, Zweier JL: High resolution electron paramagnetic resonance imaging of biological samples with a single line paramagnetic label. Magn Reson Med 1997;37: 479–483.
91 Berliner LJ, Fujii H, Wan XM, Lukiewicz SJ: Feasibility study of imaging a living murine tumor by electron paramagnetic resonance. Magn Reson Med 1987;4:380–384.
92 Bottomley PA, Andrew ER: RF magnetic field penetration, phaseshift and power dissipation in biological tissue: Implications for NMR imaging. Phys Med Biol 1978;23:630–637.
93 Mehlhorn RJ, Fuchs J, Sumida S, Packer L: Preparation of tocopheroxyl radicals for detection by electron spin resonance. Methods Enzymol 1990;186:197–205.
94 Kuppusamy P, Wang P, Shankar RA, Ma L, Trimble CE, Hsia CJ, Zweier JL: In vivo topical EPR spectroscopy and imaging of nitroxide free radicals and polynitroxyl-albumin. Magn Reson Med 1998;40:806–811.
95 Kuppusamy P, Afeworki M, Shankar RA, Coffin D, Krishna MC, Hahn SM, Mitchell JB, Zweier JL: In vivo electron paramagnetic resonance imaging of tumor heterogeneity and oxygenation in a murine model. Cancer Res 1998;58:1562–1568.
96 Colacicchi S, Ferrari M, Sotgiu A: In vivo electron paramagnetic resonance spectroscopy/imaging: First experiences, problems, and perspectives. Int J Biochem 1992;24:205–214.
97 Quaresima V, Ferrari M: Current status of electron spin resonance (ESR) for in vivo detection of free radicals. Phys Med Biol 1998;43:1937–1947.
98 Alecci M, Ferrari M, Quaresima V, Sotgiu A, Ursini CL: Simultaneous 280 MHz EPR imaging of rat organs during nitroxide free radical clearance. Biophys J 1994;67:1274–1279.

99 Halpern HJ, Peric M, Yu C, Barth ED, Chandramouli G, Makinen MW, Rosen GM: In vivo spin-label murine pharmacodynamics using low-frequency electron paramagnetic resonance imaging. Biophys J 1996;71:403–409.

100 Togashi H, Shinzawa H, Ogata T, Matsuo T, Ohno S, Saito K, Yamada N, Yokoyama H, Noda H, Oikawa K, Kamada H, Takahashi T: Spatiotemporal measurement of free radical elimination in the abdomen using an in vivo ESR-CT imaging system. Free Radic Biol Med 1998;25:1–8.

101 Hiramatsu M, Oikawa K, Noda H, Mori A, Ogata T, Kamada H: Free radical imaging by electron paramagnetic resonance computed tomography in rat brain. Brain Res 1995;30:44–47.

102 Alecci M, Lurie DJ, Nicholson I, Placidi G, Sotgiu A: A proton-electron double-resonance imaging apparatus with simultaneous multiple electron paramagnetic resonance irradiation at 10 mT. Magma 1996;4:187–193.

103 Foster MA, Seimenis I, Lurie DJ: The application of PEDRI to the study of free radicals in vivo. Phys Med Biol 1998;43:1893–1897.

104 Lurie DJ, Foster MA, Yeung D, Hutchison JM: Design, construction and use of a large-sample field-cycled PEDRI imager. Phys Med Biol 1998;43:1877–1886.

105 Yokoyama H, Sato T, Tsuchihashi N, Ogata T, Ohya-Nishiguchi H, Kamada H: A CT using longitudinally detected ESR (LODESR-CT) of intraperitoneally injected nitroxide radical in a rat's head. Magn Reson Imaging 1997;15:701–708.

106 Ohya-Nishiguchi H: Overview of bioradicals and ESR technology; in Ohya-Nishiguchi J, Packer L (eds): Bioradicals Detected by ESR Spectroscopy. Basel, Birkhäuser, 1995, pp 1–15.

107 Fuchs J, Groth N, Herrling T, Packer L: In vivo EPR skin imaging. Methods Enzymol 1994;203: 140–149.

108 Herrling T, Groth NE, Fuchs J: Biochemical EPR imaging in skin. Appl Magn Reson 1996;11: 471–486.

109 Herrling T, Zastrow L, Groth N: Classification of cosmetic products – The radical protection factor (RPF). Seifen Öle Fette Wachse J 1998;124:282–284.

110 Mäder K, Stosser R, Borchert HH: Transcutaneous absorption of nitroxide radicals detected in vivo by means of X-band ESR. Pharmazie 1992;47:946–947.

111 Mäder K, Stösser R, Borchert HH: Detection of free radicals in living mice after inhalation of DTBN by X-band ESR. Free Radic Biol Med 1993;14:339–342.

112 Lin Y, Yokoyama H, Ishida S, Tsuchihashi N, Ogata T: In vivo electron spin resonance analysis of nitroxide radicals into a rat by a flexible surface coil type resonator as an endoscope or a stethoscope like device. Magma 1997;5:99–103.

113 Mäder K, Baacic G, Swartz HM: In vivo detection of anthralin-derived free radicals in the skin of hairless mice by low-frequency electron paramagnetic resonance spectroscopy. J Invest Dermatol 1995;104:514–517.

114 Liu KJ, Mader K, Shi X, Swartz HM: Reduction of carcinogenic chromium (VI) on the skin of living rats. Magn Reson Med 1997;38:5245–5246.

115 Furusawa M, Ikeya M: Electron spin resonance imaging utilizing localized microwave magnetic field. Jpn J Appl Phys 1990;29:270–276.

116 Pryor WA, Godber SS: Noninvasive measures of oxidative stress status in humans. Free Radic Biol Med 1991;10:177–184.

Jürgen Fuchs, PhD, MD, Heinsestrasse 8, D–63739 Aschaffenburg (Germany)
Tel. +49 6021 219825, Fax +49 6021 219746, E-Mail juergenfuchs@hotmail.com

Thiele J, Elsner P (eds): Oxidants and Antioxidants in Cutaneous Biology.
Curr Probl Dermatol. Basel, Karger, 2001, vol 29, pp 18–25

Electron Paramagnetic Resonance Detection of Free Radicals in UV-Irradiated Human and Mouse Skin

Beth Anne Jurkiewicz Lange[a], *Garry R. Buettner*[b]

[a] Kimberly Clark Corporation, Neenah, Wisc., and
[b] Free Radical Research Institute, EMRB 68, University of Iowa, Iowa City, Iowa, USA

Skin is directly exposed to an oxidizing environment. Evidence suggests that signs of skin aging (wrinkling, sagging, actinic lentigines) may be due, in part, to cumulative oxidative damage incurred throughout our lifetimes [1]. UV radiation plays a role in this damage and is also commonly used experimentally to initiate oxidative damage. The hairless mouse is a widely used model for studying UV radiation damage of skin. However, there are significant limitations associated with the murine model as well as biochemical and structural differences that exist between mouse and human skin that could affect their responses to UV radiation and other environmental insults. In this chapter, we compare the standard Skh-1 hairless mouse skin to human skin biopsies as models for UV-radiation-induced free radical formation.

Skin has endogenous antioxidants to protect against free radical damage. An important water-soluble antioxidant present in the skin is ascorbate. Mice naturally produce ascorbic acid; in humans, ascorbate is an essential vitamin that must be obtained from the diet. However, in both species high levels of ascorbate are present in the epidermis. Ascorbate probably provides the major free radical sink when radicals are produced in mammalian tissues at physiological pH [2]. The ascorbate free radical ($Asc^{\bullet -}$) is detectable by electron paramagnetic resonance (EPR) spectroscopy at low steady-state levels in many diverse biological samples [3–7], including skin [8–12]. It has been demonstrated that $Asc^{\bullet -}$ can be used as a noninvasive indicator of oxidative stress [13]. As such, $Asc^{\bullet -}$ has been useful in the study of free radical oxidations in many biological systems [14–17] including mouse [8–11] and human skin.

Because $Asc^{\bullet -}$ is a resonance-stabilized free radical with a narrow line width, it is easily detectable directly by EPR. However, the non-resonance-stabilized free radicals initially produced in the skin by UV radiation have very short lifetimes at room temperature; thus, EPR spin trapping techniques must be applied to detect these radicals.

Materials and Methods

A common approach for studying free radicals produced in skin during UV radiation exposure has been to use homogenized skin samples [18, 19]. Homogenizing the sample will produce artifacts; thus, in this study intact epidermal sections of mouse or human skin are used to study real-time free radical formation by UV radiation. For animal experiments, whole dorsal skin was harvested from Skh-1 hairless mice (Charles River Laboratory, Portage, Mich., USA) and cut into EPR-usable pieces (≈ 1 cm^2). For human studies, skin sections were obtained from pre- or postauricular regions and were immediately frozen at 77 K. Because of limitations in EPR sample size, much of the dermis was cut from the human skin samples prior to examination.

The epidermal surface of the skin was exposed to UV radiation while the skin sample was in the EPR cavity. It was assumed that the EPR cavity grid transmits 75% of the incident radiation. The light source was an Oriel 150 W Photomax® xenon arc lamp operating at about 30 W power. The filtered light fluence rate at the sample was estimated using a Model IL 1400A International Light Inc. radiometer with UVA detector, model SEL 033; the UVB detector was a model SEL 240. Wavelengths below ≈ 300 nm were filtered out using a Schott WG 305 filter (14 $\mu W/cm^2$ UVB; 3.5 mW/cm^2 UVA). For visible light experiments, wavelengths below 400 nm were filtered out using an Oriel 59472 filter (0.23 mW/cm^2 UVA). Infrared radiation was removed in all experiments by a 5.0-cm quartz water filter.

A Bruker ESP-300 EPR spectrometer (Karlsruhe, Germany) was used for all experiments. The settings for the ascorbate radical experiments were: microwave power, 40 mW; modulation amplitude, 0.66 GG; time constant, 0.3 s; scan rate, 8 GG/41.9 s; receiver gain, 2×10^6. The EPR instrument settings for the α-(4-pyridyl-1-oxide)-N-tert-butylnitrone (POBN) spin trapping experiments were: microwave power, 40 mW; modulation amplitude, 0.76 GG; time constant, 0.3 s; scan rate, 60 GG/41.9 s; receiver gain, 1×10^6. The EPR spectrometer setting for the 5,5-dimethylpyrroline-1-oxide (DMPO) experiments were: microwave power, 40 mW; modulation amplitude, 1.06 GG; time constant, 0.3 s; scan rate, 80 GG/84 s; receiver gain, 1×10^6. For the spin trapping experiments, no increase in background EPR signal occurred when an aqueous solution of either POBN or DMPO was exposed to UV radiation at the doses used in the skin experiments.

Results

At room temperature, endogenous, resonance-stabilized Asc^{-} are detectable by EPR at a low steady-state level in mouse (Skh-1) skin [8, 9] and in human skin biopsies ($g = 2.0053$, $a^{H4} \cong 1.8$ GG). Exposure of mouse skin,

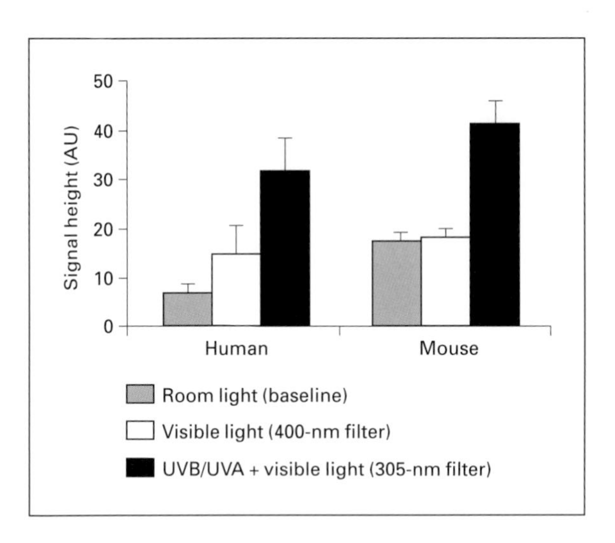

Fig. 1. Asc$^{\bullet -}$ increases in mouse and human skin when exposed to UV radiation. The data expressed in arbitrary units (AU) represent the means of 4 separate experiments. The skin was exposed to UV radiation or visible light after collection of a steady-state baseline reading. For mouse skin, after light exposure, a paired comparison of the data showed that the visible-light-exposed samples were not to be statistically different from those samples exposed to room light. For the human skin, there was a significant increase in the signal obtained in room light versus visible light ($p < 0.05$). UV radiation significantly enhanced radical signal height in both species.

while in the EPR cavity, to a combined UVA and UVB radiation source results in an approximately twofold increase in the Asc$^{\bullet -}$ signal height relative to unirradiated mouse skin, indicating that during UV radiation exposure of the skin is undergoing free radical oxidative stress (fig. 1). However, when human skin is exposed to the same UV radiation source there is an approximate fourfold increase in the Asc$^{\bullet -}$ EPR signal intensity. The kinetics of the observed changes in Asc$^{\bullet -}$ EPR signal height between human and mouse skin also differed, as did the patterns of spontaneous Asc$^{\bullet -}$ radical accumulation in nonirradiated skin (data not shown). The differences in radical signal response may be due to endogenous skin ascorbate levels or due to variation in skin lipid composition and consequently target for free radical damage between the two species.

Visible light photons have generally been considered innocuous because of their low energies; however, in skin, naturally occurring photosensitizers could produce free radicals in the presence of visible light. Thus, the potential involvement of visible light in free radical production was examined. Exposure of mouse skin to visible light, i.e. wavelengths above ≈400 nm, does not increase the $Asc^{\bullet -}$ signal above ambient light levels; indicating that visible light has no detectable effect on free radical formation in mouse skin.

However, in human skin, visible light was found to increase $Asc^{\bullet -}$ approximately twofold. In human skin there may be visible light chromophore not found in Skh-1 mouse skin, such as metals, flavins and melanin, which can produce free radicals. It has been shown that 694-nm ruby laser light did not result in free radical formation in human skin, indicating that the chromophores may be wavelength specific [20].

Spin trapping techniques were applied to further identify the short-lived initial free radicals produced by UV radiation in murine skin. Using the spin trap POBN, a carbon-centered POBN spin adduct as well as $Asc^{\bullet -}$ is observable in UV-radiation-exposed mouse skin. The spectra exhibit hyperfine splittings characteristic of POBN/alkyl radicals, $a^N = 15.56$ GG and $a^H = 2.70$ GG, possibly from membrane lipids as a result of β-scission of lipid alkoxyl radicals causing generation of alkyl radicals, such as ethyl and pentyl radicals [21, 22]. DMPO was also used to further identify the radicals involved in UV-radiation-induced free radical formation in mouse skin. However, no spin adduct was observable in the absence or presence of UV radiation. Only the ascorbate radical was observed.

When POBN was topically applied to the human skin biopsies, no EPR signal was observable in the absence of UV radiation. Exposure to UV radiation resulted in a triplet of doublets as well as the $Asc^{\bullet -}$ signal. The hyperfine splittings of the POBN adduct were $a^N = 15.62$ GG and $a^H = 3.1$ GG, possibly indicating the trapping of the carbon dioxide radical anion [22, 23]. This signal was very weak, thus DMPO was also explored as a spin trap. When human skin biopsies were treated with DMPO, no EPR signal was detected in room light alone. Upon UV radiation exposure, $Asc^{\bullet -}$ as well as a 6-line DMPO adduct signal (hyperfine splittings, $a^N = 14.54$ GG and $a^H = 16.0$ GG) were observed, characteristic of alkoxyl radicals [11, 22, 24] possibly formed by lipid peroxidation processes. The DMPO spin adduct splittings in the human skin biopsies are also similar to those of $DMPO/SO_3^{\bullet -}$; however, a lipid-derived alkoxyl radical is more likely in this system. This presence of lipid alkoxyl radicals in human skin is consistent with the observation of UV-radiation-induced lipid alkyl radicals in mouse skin [9, 10].

As lipid peroxidation products were indicated in our human skin spin trapping experiments, iron may be involved in these free radical processes. Desferal®, an iron-chelating agent, was examined as a protectant against UV-radiation-induced free radical production. Desferal was topically applied to the human skin prior to EPR examination. Using DMPO spin trapping techniques, after exposure to UV radiation the same 6-line DMPO adduct, as well as

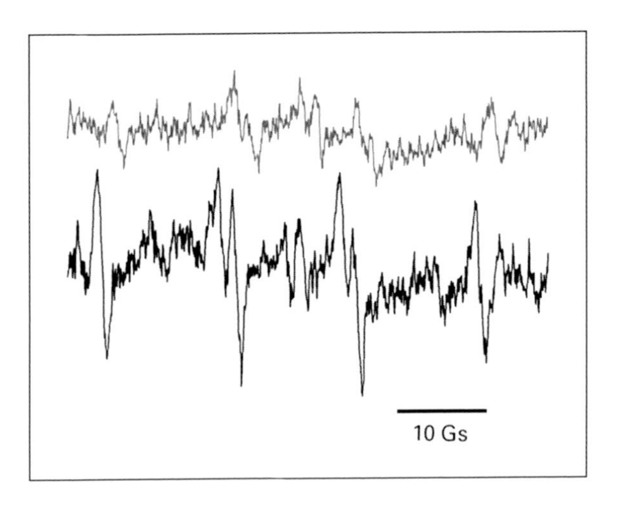

Fig. 2. Desferal decreases the UV-radiation-induced DMPO adduct signal in human skin biopsies. The lower spectrum is of the DMPO signal adduct from human skin exposed to UV radiation. The upper spectrum is of human skin topically treated with 100 µl of 10 m*M* Desferal for 10 min followed by exposure to UV radiation. Both spectra are the result of 5 signal-averaged scans.

$Asc^{\bullet -}$, was observable in both treatment groups. Desferal reduced the DMPO radical signal by ≈50% (fig. 2), which parallels earlier observations in mouse skin [9].

Discussion

It has been demonstrated that the ascorbate radical EPR signal can serve as an indicator of oxidative events [6, 9, 13]. In both mouse and human skin biopsies, the $Asc^{\bullet -}$ signal intensity increases with UV radiation exposure, indicating oxidative stress. Visible light also causes a modest increase in free radical formation in the human biopsies, in contrast to mouse skin. Direct detection of $Asc^{\bullet -}$ by EPR is consistent with ascorbate's role as the terminal small-molecule antioxidant. Due to the ascorbate radical's low reduction potential, +282 $E^{o\prime}$ mV [25], nearly every oxidizing radical that can arise in a biological system will react with ascorbate forming the semi-dehydroascorbate radical ($Asc^{\bullet -}$, a resonance-stabilized, tricarbonyl free radical species. Thus, UV light increases the flux of radicals in both mouse and human skin.

Exposure of mouse skin to UVB radiation or benzoyl peroxide produces oxidative changes in skin that qualitatively resemble each other, suggesting that common mechanisms are involved [26]. Indeed, organic peroxides induce radicals in mouse skin characteristic of trapping peroxyl, alkoxyl and alkyl radicals [11]. These species are consistent with radicals induced by UV radiation in skin.

Spin-trapped radicals in human biopsies appear to be different from the radicals trapped from mouse skin, i.e. an alkoxyl versus an alkyl radical [9, 27]. The presence of lipid peroxidation products in the mouse and human skin models is not surprising considering that in normal human skin the content of lipid peroxides is higher in sun-exposed sites than in non-sun-exposed sites [28], and in chronically exposed areas of human skin lipid peroxidation products are increased following UV exposure [29].

Iron may also play a role in oxidative skin damage. In this work, the metal chelator Desferal was found to reduce free radical formation in skin indicating that iron may play a role in the observed UV-induced oxidative damage. UVA radiation of human skin cells induces release of iron [30], a known contributor to free radical reactions. In addition, chronic UV-exposed mouse and human skin samples have higher iron levels. The presence of this auxiliary iron may cause even more deleterious oxidative reactions in already damaged skin. Application of metal chelators may be a route to prevent or reduce oxidative damage in skin.

Vitamin E is also beneficial in protecting lipid membranes against peroxidation. Topical application of α-tocopherol or tocopherol sorbate to mouse skin is effective in reducing basal as well as UV-radiation-induced free radical formation [10]. This reduction in free radical levels appears to correlate with a reduction in skin wrinkling due to chronic UV radiation exposure. Vitamin E and its analogs may provide topical benefits in human skin by reducing oxidative damage [31, 32].

Despite the differences between mouse and human skin, the mouse is a useful model for examining UV-radiation-induced free radical events in skin. Understanding deleterious oxidation events in skin will lead to the development of technologies such as metal chelators and antioxidant treatments to mitigate free radical events prior to inception of dermatopathological changes.

References

1 Miyachi Y: Photoaging from an oxidative standpoint. J Dermatol Sci 1995;9:79–86.
2 Wardman P: Evaluation of the 'radical sink' hypothesis from a chemical-kinetic viewpoint. J Radioanal Nucl Chem 1998;232:23–27.
3 Stegmann HB, Schuler P, Westphal S, Wagner E: Oxidative stress of crops monitored by EPR. Z Naturforsch C 1993;48:766–772.

4 Rose RC, Bode AM: Biology of free radical scavengers: An evaluation of ascorbate. FASEB J 1993; 7:1135–1142.
5 Minetti M, Forte T, Soriani M, Quaresima V, Mendito A, Ferrari M: Iron-induced ascorbate oxidation in plasma as monitored by ascorbate free radical formation. Biochem J 1992;282:459–465.
6 Sharma MK, Buettner GR: Interaction of vitamin C and vitamin E during free radical stress in plasma: An ESR study. Free Radic Biol Med 1993;14:649–653.
7 Buettner GR, Chamulitrat W: The catalytic activity of iron in synovial fluid as monitored by the ascorbate free radical. Free Radic Biol Med 1990;8:55–56.
8 Buettner GR, Motten AG, Hall RD, Chignell CF: ESR detection of endogenous ascorbate free radical in mouse skin: Enhancement of radical production during UV irradiation following topical application of chlorpromazine. Photochem Photobiol 1987;46:161–164.
9 Jurkiewicz BA, Buettner GR: Ultraviolet light-induced free radical formation in skin: An electron paramagnetic resonance study. Photochem Photobiol 1994;59:1–4.
10 Jurkiewicz BA, Bissett DL, Buettner GR: The effect of topically applied tocopherol on ultraviolet-radiation mediated free radical damage in skin. J Invest Dermatol 1995;104:484–488.
11 Timmins GS, Davies MJ: Free radical formation in murine skin treated with tumour promoting organic peroxides. Carcinogenesis 1993;14:1499–1503.
12 Kitazawa M, Podda M, Thiele J, Traber MG, Iwasaki K, Sakamoto K, Packer L: Interactions between vitamin E homologues and ascorbate free radicals in murine skin homogenates irradiated with ultraviolet light. Photochem Photobiol 1997;65:355–365.
13 Buettner GR, Jurkiewicz BA: Ascorbate free radical as a marker of oxidative stress: An EPR study. Free Radic Biol Med 1993;14:49–55.
14 Tomasi A, Albano E, Bini A, Iannone AC, Vannini V: Ascorbyl radical is detected in rat isolated hepatocytes suspensions undergoing oxidative stress: An early index of oxidative damage in cells. Adv Biosci 1989;76:325–334.
15 Arroyo CM, Kramer JH, Dickens BF, Weglicki WB: Identification of free radicals in myocardial ischemia/reperfusion by spin trapping with nitrone DMPO. FEBS Lett 1987;221:101–104.
16 Nohl H, Stolze K, Napetschnig S, Ishikawa T: Is oxidative stress primarily involved in reperfusion injury of the ischemic heart? Free Radic Biol Med 1991;11:581–588.
17 Sharma MK, Buettner GR, Spencer KT, Kerber RE: Ascorbyl free radical as a real-time marker of free radical generation in briefly ischemic and reperfused hearts. Circ Res 1994;74:650–658.
18 Ogura R, Sugiyama M, Nishi J, Haramaki N: Mechanism of lipid radical formation following exposure of epidermal homogenate to ultraviolet light. J Invest Dermatol 1991;97:1044–1047.
19 Ogura R, Nishi J, Sugiyama M, Haramaki N, Kotegawa M: Lipid radical formation of the epidermis exposed to ultraviolet light (ESR study). Photomed Photobiol 1990;12:179–184.
20 Haywood RM, Wardman P, Gault DT, Linge C: Ruby laser irradiation (694 nm) of human biopsies: Assessment by electron spin resonance spectroscopy of free radical production and oxidative stress during laser depilation. Photochem Photobiol 1999;70:348–352.
21 North JA, Spector AA, Buettner GR: Detection of lipid radicals by electron paramagnetic resonance spin trapping using intact cells enriched with polyunsaturated fatty acid. J Biol Chem 1992;267: 5743–5746.
22 Buettner GR: Spin trapping: ESR parameters of spin adducts. Free Radic Biol Med 1987;3:259–303.
23 Li ASW, Cummings KB, Roethling HP, Buettner GR, Chignell CF: A spin trapping database implemented on the IBM PC/AT. J Magn Res 1988;19:140–142.
24 Smith FL, Floyd RA, Carpenter MP: Prostaglandin synthase-dependent spin trapped free radicals; in Rodgers MAJ, Powers EL (eds): Oxygen and Oxy-Radicals in Chemistry and Biology. New York, Academic Press, 1981, pp 734–745.
25 Williams NH, Yandell JK: Reduction of oxidized cytochrome c by ascorbate ion. Biochim Biophys Acta 1985;810:274–277.
26 Ibbotson SH, Moran MN, Nash JF, Kochevar IE: The effects of radicals compared with UVB as initiating species for the induction of chronic cutaneous photodamage. J Invest Dermatol 1999;112: 933–938.
27 Jurkiewicz BA, Buettner GR: EPR detection of free radicals in UV-irradiated skin: Mouse versus human. Photochem Photobiol 1996;64:918–922.

28 Bissett DL, McBride JF: Iron content of human epidermis from sun-exposed and non-exposed body sites. J Soc Cosmet Chem 1992;43:215–217.
29 Meffert H, Diezel W, Sonnichsen N: Stable lipid peroxidation products in human skin: Detection, ultraviolet light-induced increase, pathogenic importance. Experientia 1976;32:1397–1398.
30 Pourzand C, Watkin RD, Brown JE, Tyrrell RM: Ultraviolet A radiation induces immediate release of iron in human primary skin fibroblasts: The role of ferritin. Proc Natl Acad Sci USA 1999;96: 6751–6756.
31 Dreher F, Gabard B, Schwindt DA, Maibach HI: Topical melatonin in combination with vitamins E and C protects skin from ultraviolet-induced erythema: A human study in vivo. Br J Dermatol 1998;139:332–339.
32 Keller KL, Fenske NA: Uses of vitamins A, C, and E and related compounds in dermatology: A review. J Am Acad Dermatol 1998;39:611–625.

Dr. Beth Anne Jurkiewicz Lange, Kimberly Clark Corporation,
2100 Winchester Road, Neenah, WI 54956 (USA)
Tel. +1 920 721 5096, Fax +1 920 721 7929, E-Mail balange@kcc.com

Thiele J, Elsner P (eds): Oxidants and Antioxidants in Cutaneous Biology.
Curr Probl Dermatol. Basel, Karger, 2001, vol 29, pp 26–42

The Antioxidant Network of the Stratum corneum

Jens J. Thiele[a], *Christina Schroeter*[a], *Sherry N. Hsieh*[a], *Maurizio Podda*[b], *Lester Packer*[c]

[a] Department of Dermatology and Allergology, Friedrich Schiller University, Jena, and
[b] Department of Dermatology, J.W. Goethe University, Frankfurt, Germany;
[c] Department of Molecular and Cell Biology, University of California, Berkeley, Calif., USA

Located at the interface between body and environment, the stratum corneum (SC) is frequently and directly exposed to a prooxidative environment, including air pollutants, UV solar light, chemical oxidants and microorganisms [1, 2]. To counteract oxidative injury of structural lipids and proteins, human skin is equipped with a network of enzymatic and nonenzymatic antioxidant systems [3]. The SC is comprised of a unique, highly lipophilic 2-compartment system of structural, enucleated cells (corneocytes) embedded in a lipid-enriched intercellular matrix, forming stacks of bilayers that are rich in ceramides, cholesterol and free fatty acids [4]. The SC lipid composition and structure plays a key role in determining barrier integrity, which is essential for skin moisturization, normal desquamation and healthy skin condition [5].

While many unique histological, biophysical and biochemical features of the SC have been known for long, its redox properties have only recently become the subject of systematic basic research (fig. 1). The aim of this chapter is to provide a brief overview on the relevant data available on this emerging field of research.

Physiological Levels of Antioxidants in the Stratum corneum

Vitamin E

α-Tocopherol, the major biologically active vitamin E homologue, is generally regarded as the most important lipid-soluble antioxidant in human tissues [6].

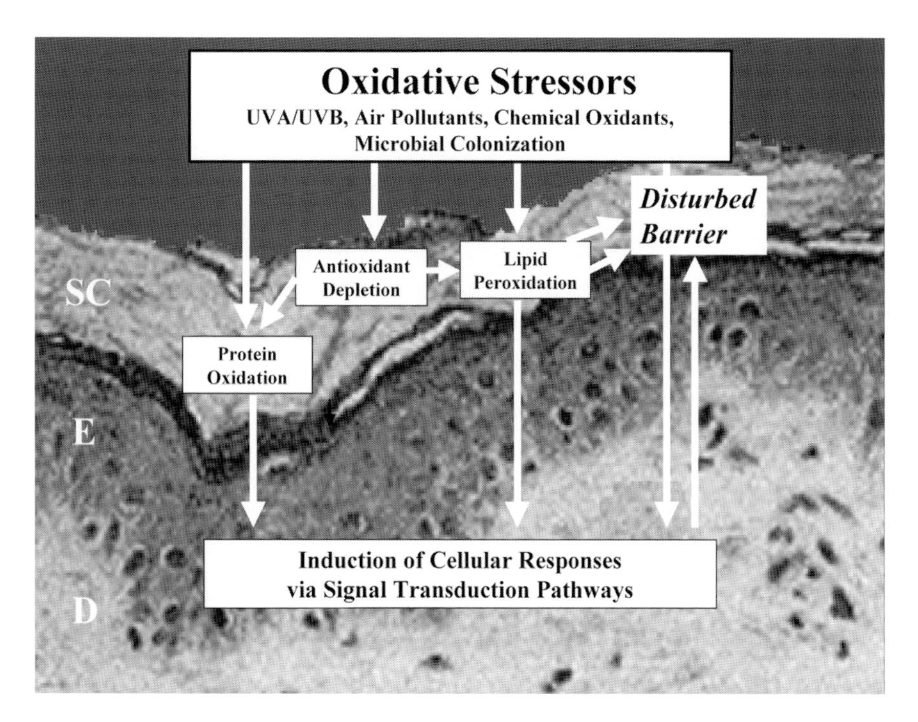

Fig. 1. Hypothetical scheme of oxidative stress in skin: effect of different oxidative stressors on the skin barrier and induction of cellular responses. E = Epidermis; D = dermis.

Remarkably, a vitamin E gradient was found in the SC of untreated, healthy human skin, with the lowest tocopherol concentrations on the surface and the highest in the deepest SC layers. In human epidermis, the ratio of α- and γ-tocopherol is about 10:1 [7]. This is in accordance with the α- and γ-tocopherol ratio we have found in the deepest SC layer, with 10-fold higher α-tocopherol compared with γ-tocopherol concentrations. The in vivo antioxidant activity of α-tocopherol is thought to be higher than that of γ-tocopherol.

Besides its protection against lipid peroxidation, vitamin E is suggested to stabilize the lipid bilayer, which may also be of relevance for SC lipid bilayers: the degree of disorder and the amount of lipids decrease over the outer cell layers of human SC [8]. Thus, low levels of SC vitamin E are associated with a high degree of SC lipid disorder.

Vitamin C, Glutathione and Uric Acid

In a recent study, we have established the physiological baseline concentrations of hydrophilic antioxidants in healthy human upper arm skin [9].

Remarkably, the concentrations of ascorbate detected in human SC were lowest in the outer SC and increasing almost 10-fold in the lower SC, however still between 1 and 2 orders of magnitude lower than epidermal ascorbate concentrations. Urate concentrations within human SC were distributed more evenly than ascorbate and tocopherols; its concentration in the upper SC was more than 100-fold higher than ascorbate, and a small increase towards deeper layers was only significant when expressed per milligram wet weight, not when expressed per extracted protein. Similar gradients with highest levels in basal SC layers were found for ascorbate, glutathione and uric acid in the SC of the untreated hairless mouse [10].

Enzymatic Antioxidants in the Stratum corneum

Only very little published information is available on the presence of enzymatic antioxidants in the SC. Previous studies reported a lack of significant activities of glutathione peroxidase, catalase and superoxide dismutase in tape-stripped human SC [3]. However, this may have been due to methodological problems; other studies reported the presence of catalase in upper SC layers of healthy human skin, while the SC of patients with polymorphic light eruption contained significantly decreased catalase levels [11]. Catalase activity in human SC depends on the anatomical site, follows a gradient with the lowest levels in the uppermost layers and is depleted by UVA irradiation [pers. commun. Dr. Nguyen, L'Oréal, Clichy, France].

Impact of Environmental Factors on Barrier-Antioxidants

Ozone

Recently, we have published a series of studies investigating the effects of the air pollutant O_3 on antioxidants and lipids in skin, which have provoked an extensive interest in SC α-tocopherol and other SC antioxidants [1, 10, 12–14]. These studies systematically led to the identification of the SC as the skin layer most susceptible to oxidative ozone damage. Even though the relevance of ozone-induced oxidative stress for human skin remains unclear, these studies were the first to demonstrate that, in principle, the SC is susceptible to oxidative attack by atmospheric oxidants. Since these studies have initiated the new field of redox research in the skin barrier, we have dedicated a separate chapter to this topic in this book [15].

UVA and UVB

Photooxidative Stress in Skin

Exposure of mammalian skin to UV irradiation induces a spectrum of well-documented acute and chronic responses, including erythema, hyperproliferation, desquamation and permeability barrier alterations. A diminished permeability barrier function has been demonstrated after single exposures to UVA and UVB [16]. However, the underlying mechanisms of UV-radiation-induced changes in SC barrier function remain unclear.

Since it is well known that UVA and UVB irradiation induces the formation of reactive oxygen species (ROS) in cutaneous tissues [17], numerous studies have focused on establishing baseline levels of antioxidants in the dermis and epidermis [7], the antioxidant response to UVA and UVB light in these, and the evaluation of photoprotective potential of topical antioxidant supplementation.

A prime mechanism of UVA- and UVB-induced damage to cutaneous tissues is thought to be the peroxidation of lipids. Vitamin E has been demonstrated to provide photoprotective effects in cell culture and in hairless mouse skin [3].

Effect on Vitamin E and Lipid Peroxidation

In a recent study we demonstrated that SC antioxidants are dramatically depleted by solar-simulated UV (SSUV) irradiation already at very low doses: a single suberythemogenic dose of SSUV light (0.75 minimal erythemal dose, MED) depleted human SC α-tocopherol by almost 50% and murine SC α-tocopherol by 85% [18]. In previous studies, SSUV doses equivalent to 3 MED or more were necessary to detect a significant depletion of α-tocopherol in whole epidermis and dermis [19–21]. The high susceptibility of SC vitamin E to SSUV may be, at least in part, due to a lack of co-antioxidants in the SC. In vitro, ubiquinol 10 protects α-tocopherol from photooxidation by recycling mechanisms [22]. In SSUV-irradiated murine skin homogenates, ascorbate, the major hydrophilic co-antioxidant, can recycle photooxidized α-tocopherol [16] (fig. 2). However, in murine and human SC, the levels of ascorbate were very low as compared to epidermal and dermal tissue, which is not surprising since the SC is a very hydrophobic tissue. While α-tocopherol in murine SC was significantly depleted after suberythemogenic UV irradiation, the lipid peroxidation parameter malondialdehyde was increased only after an unphysiologically high dose. Similarly, in previous studies on the cutaneous effects of ozone we found that the dose necessary to detect increased malondialdehyde formation in the SC was 5 times higher than the one to deplete vitamin E [14].

In theory, the observed SC tocopherol depletion could be caused either directly by absorption of short-wave UVB and/or indirectly by excited-state

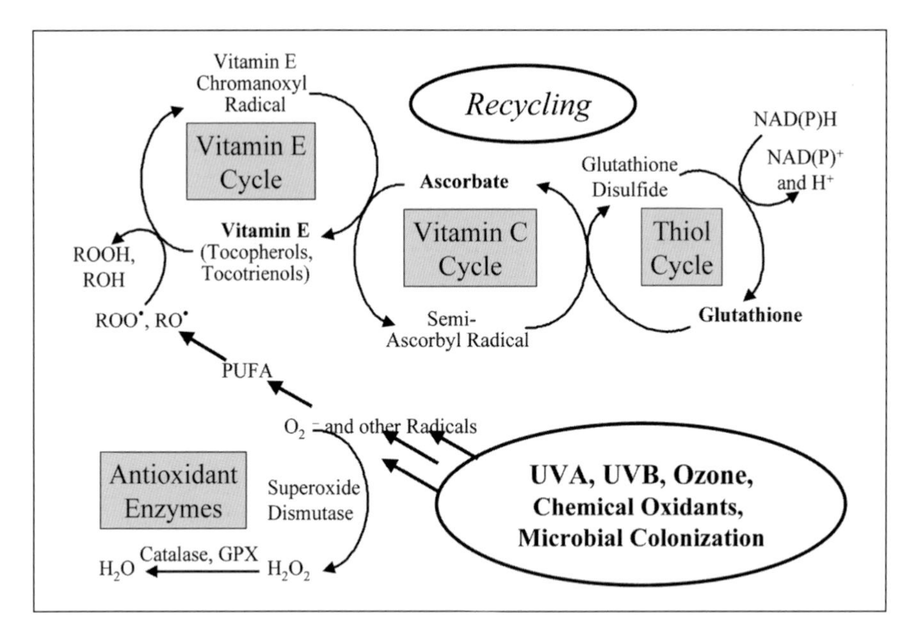

Fig. 2. The antioxidant network in skin: predominant role of α-tocopherol and its recycling by co-antioxidants. GPX = Glutathione peroxidase.

singlet oxygen or reactive oxygen intermediates that are generated by photosensitizers upon UV absorption also in the UVA range. Our results indicate that both mechanisms may be relevant, since either UVB or UVA depleted murine SC α-tocopherol. The absorption maxima of both tocopherols are very close together, which suggests that mechanisms other than direct photodestruction by UV, e.g. free radical scavenging properties, may account for this difference in their susceptibilities to UV-mediated depletion.

Benzoyl Peroxide

Benzoyl peroxide (BP) is widely used in the treatment of acne. Since BP is known to be a strong oxidant and to decompose within the SC, we hypothesized that BP affects the antioxidant defense capacity in the SC. To test this, we sought to evaluate the impact of 10%-BP-treated versus vehicle-treated skin on SC antioxidants, and to compare the antioxidant response with conventional biophysical measurements of SC function.

New methods were developed to detect SC concentrations of urate, vitamin C and vitamin E, based on electrochemical HPLC analysis of SC extracts from sequential tape strippings. Baseline antioxidant levels were evaluated in 24 consecutive tape strippings of upper arm SC (n = 12). BP or vehicle (contralateral site) were applied to upper arm SC in 9 volunteers, and after 22 h the transepidermal water loss was measured. Then, SC antioxidants were analyzed (strippings No. 1–8: 'upper SC', No. 9–16: 'intermediate SC', and No. 17–24: 'lower SC'). In this study, BP treatment dramatically depleted vitamin E by 95% in the upper and intermediate SC, and by 85% in the lower SC (both $p<0.001$); vitamin C was reduced by 65% in the intermediate and lower SC ($p<0.001$), whereas urate depletion was strongest in the upper SC (by 70%, $p<0.001$). Notably, the transepidermal water loss and redness did not differ in untreated, vehicle-treated or BP-treated skin.

BP was used as a model compound for radical-generating chemicals/drugs in order to investigate its impact on SC antioxidants. Furthermore, since BP is known to be a moderate skin irritant, we wanted to compare the effect on SC antioxidants with conventional bioengineering methods used to assess SC integrity. Although the vehicle used here, acetone (control), itself can remove SC lipids and thus affect the SC integrity, under the conditions used in the present study, no increase in the transepidermal water loss and no erythema was visible 1 day after the 10% BP in acetone or the acetone application. The hydration was slightly lower in the BP-treated skin as compared to untreated, but not as compared to control (acetone-treated) skin sites. In contrast, our new methods for SC antioxidant detection revealed dramatic changes: While the hydrophilic antioxidants urate and ascorbate showed differing susceptibility in the upper and lower SC, α-tocopherol was most dramatically depleted by more than 90% throughout the entire SC thickness.

We concluded that SC antioxidant depletion, particularly in the case of α-tocopherol, represents an early and very sensitive biomarker not only for ozone- and UV-induced oxidation processes in the barrier, but also for chemical/drug exposure. It has to be pointed out that the conditions used in this study were (a) in vivo, (b) in humans and (c) using a relevant BP concentration as it is frequently used. Thus, it is very likely that barrier α-tocopherol depletion occurs during BP use in acne treatment. Vitamin E depletion may account, at least in part, for common side effects such as dry and scaly skin. Moreover, these observations are in favor of our hypothesis that redox processes in the SC play a role for desquamatory processes in skin (see below, 'Desquamation').

Physiological Mechanisms of Barrier Antioxidant Repletion

Regional Differences in Vitamin E and Sebum Levels

The first studies on sebum antioxidants were based on the rather unexpected observation that the upper SC layers of the environmentally exposed human facial skin contained severalfold higher levels of α-tocopherol than corresponding layers of the previously investigated, less exposed upper arm SC [18]. It was suggested that this finding may be related to regional differences in the delivery pathway and/or regulation of vitamin E. In the environmentally highly exposed facial skin, the stratum corneum is covered by a film of skin surface lipids which consist of wax esters, triglycerides and squalene, originating from sebum secretion by sebaceous glands. We hypothesized that sebaceous gland secretion is a physiological pathway of vitamin E delivery to the skin surface lipids and SC of facial skin. Therefore, we sought to examine (1) SC levels and distribution profiles of vitamin E in sites with different sebaceous gland density, (2) whether vitamin E is secreted by sebaceous glands and (3) whether there is a correlation between levels of vitamin E and squalene in skin surface lipids.

Human Sebum Contains High Amounts of Vitamin E

Using standardized techniques for SC tape stripping and sebum collection, followed by HPLC analysis of tocopherols and squalene, we found that (a) the ratio of cheek versus upper arm α-tocopherol levels was 20:1 for the upper SC and decreased gradually with SC depth, (b) vitamin E (α- and γ-tocopherol forms) is a significant constituent of human sebum and is continuously secreted at cheek and forehead sites during a test period of 135 min and (c) vitamin E correlates very well with levels of co-secreted squalene ($r^2=0.86$, $p<0.001$).

Intriguingly, while a large body of evidence points to photoprotective effects of topically applied vitamin E against immunosuppression, DNA damage and carcinogenesis, little is known about the role of physiological vitamin E regulation in cutaneous tissues. Remarkably, the herein reported human sebum levels of α-tocopherol (moles per wet weight) are more than 3-fold higher than levels found in human blood plasma [23], human dermis, epidermis [7] and SC [18]. Sebaceous α-tocopherol most likely penetrates into subjacent SC layers, as was demonstrated for other sebum lipids [24]. Thus, it would account for the increased levels of α-tocopherol detected in the upper SC of the sebaceous gland regions of facial skin as compared to upper arm skin. These findings suggest that sebaceous gland secretion is a relevant physiological delivery pathway of α-tocopherol to sebaceous-gland-rich skin regions, such as the environmentally exposed facial skin. Similarly, orally administered drugs have been reported to be transported to the skin surface and the SC by the sebaceous gland secretion route [25].

Vitamin E Inhibits UV-Induced Squalene Oxidation

The mechanism by which α-tocopherol protects lipids against oxidative damage involves the reaction of its phenolic hydroxyl group with an organic peroxyl radical, thereby effectively interrupting the radical chain of lipid peroxidation reactions [6]. Since the levels of α-tocopherol in sebum are substantially higher than those reported for skin and most other body tissues and fluids [26], the question arises whether this serves a physiological defense strategy to protect specific sebum lipids. A prime candidate appears to be squalene, a major sebum constituent of predominantly sebaceous origin. Nonoxidized squalene was reported to be an efficient quencher of singlet oxygen in vitro and to inhibit skin-irritant-induced erythema formation in rats, as well as the activity of tumor-promoting agents in mouse skin carcinogenesis in vivo. However, upon higher oxidative challenges with singlet oxygen, squalene is readily oxidized to mono-, di- and trihydroperoxides, while unsaturated phospholipids are more stable. In skin, singlet oxygen is produced upon UVA exposures of endogenous and exogenous photosensitizers, such as psoralens and chlorpromazine, and leads to decomposition of squalene and formation of squalene peroxides. Among the various human skin surface lipids, squalene was shown to be the most susceptible to UVB-induced photooxidation [27]. Furthermore, it has been demonstrated that both UVB and UVA irradiation of squalene yield squalene peroxides and malondialdehyde, a stable, cytotoxic lipid peroxidation product. Notably, there is evidence that squalene oxidation products induce a number of harmful effects in skin cell cultures and in vivo. Low levels of UVB-peroxidated squalene increased the rates of DNA and protein synthesis in human keratinocytes, while high levels were cytotoxic [27]. In the same study, similar histological changes were reported in guinea pig skin treated with UVB-peroxidated squalene as well as with direct UVB irradiation. Furthermore, similar to UVB-induced immunosuppression, topically applied UVB-peroxidated squalene was shown to inhibit the induction of contact hypersensitivity to dinitrofluorobenzene in mice. In the rabbit ear model, UVA-peroxidated squalene was highly comedogenic, while unoxidized squalene was scarcely comedogenic, and similar effects were demonstrated in humans [for a complete reference list of reviewed studies on redox aspects of squalene and UV in skin, see 28].

Other Mechanisms of Vitamin E Regulation

Other mechanisms of passive vitamin E regulation in skin may involve epidermal renewal and differentiation processes that lead to a gradual transportation of epidermal membrane-bound lipids including vitamin E. This movement alone would, however, not explain the vitamin E gradients found in facial skin with high amounts of vitamin E in the upper SC layers. Actively regulated pathways of epidermal vitamin E are currently not known. Recent experiments carried out by our own group point to the existence of an α-tocopherol transfer protein, which was demonstrated to regulate vitamin E plasma levels and so far was only found in liver tissue [6]. However, the existence of an α-tocopherol transfer protein in cutaneous tissues needs further confirmation and should therefore not yet be considered a relevant regulation mechanism in skin.

Conclusion

In summary, our data provide conclusive evidence that sebaceous gland secretion is a relevant physiological delivery route for vitamin E to sebaceous-gland-rich skin. In view of the adverse effects of squalene peroxides on skin and the powerful antioxidative potential of vitamin E, the herein presented correlation between α-tocopherol and squalene levels in human sebum may reflect a physiological antioxidant strategy to maintain low levels of squalene oxidation products in skin surface lipids and their penetration into subjacent skin layers.

Clinical Implications of Oxidative Stress in the Stratum corneum

Photoprotection

The skin is equipped with at least two barriers to UV radiation: a melanin barrier in the epidermis and a protein barrier in the SC. The SC is a major optically protective element of the epidermis, reflecting approximately 5% of the incident light and absorbing significant portions of UVC (200–280 nm) and UVB (280–320 nm). With respect to erythema formation, the SC thickness was shown to be a main photoprotective factor, even more relevant than pigmentation and the thickness of viable epidermis [29]. Only few studies on the mechanisms of action of the photoprotective properties of the SC are currently available. Recent studies by us and other groups on redox properties of the SC, which are reviewed in this chapter, point to a relevance of UV-induced oxidative changes in the barrier and the protective action of SC antioxidants. Topically applied antioxidants provide protection against UVB-induced oxidative damage in SC lipids [30]. Even some systemically applied antioxidants accumulate in the SC and play an important role against UV-induced photodamage in skin. Intriguingly, studies on the photoprotective mechanisms of the antioxidant butylated hydroxytoluene suggested that changes in the physicochemical properties of SC keratins occurred, leading to increases in UV absorption of underlying epidermal layers. These changes were proposed to be exerted via the antiradical action of butylated hydroxytoluene that retards oxidation and prevents cross-linking of the keratin chains, resulting in a diminution of UVB radiation reaching potential epidermal target sites [31]. Similar mechanisms of action are currently discussed for other lipophilic antioxidants like tocopherols and carotenoids [3].

In conclusion, there is growing experimental evidence that cutaneous photoprotection should involve antioxidant strategies to prevent damage to SC lipids and proteins.

Desquamation

While a number of studies have demonstrated that SC lipids and lipophilic antioxidants are highly susceptible to environmentally induced oxidative stress, little is known about protein oxidation in the SC.

Oxidative Protein Modifications: General Mechanisms

Proteins are known as important targets for oxidative modifications. Oxygen radicals and other activated oxygen species generated as by-products of cellular metabolism or from environmental sources cause modifications of the amino acids of proteins that generally result in functional changes in structural or enzymatic proteins. In addition to the modification of amino acid side chains, oxidation reactions can also mediate fragmentation of polypeptide chains and both intra- and intermolecular cross-linking of peptides and proteins [32]. Protein carbonyls may be formed either by oxidative cleavage of proteins or by direct oxidation of lysine, arginine, proline and threonine residues. In addition, carbonyl groups may be introduced into proteins by reactions with aldehydes (4-hydroxy-2-nonenal, malondialdehyde) produced dur- ing lipid peroxidation or with reactive carbonyl derivatives generated as a consequence of the reaction of reducing sugars or their oxidation products with lysine residues of proteins [33]. The presence of carbonyl groups in proteins has therefore been used as a marker of reactive-oxygen-mediated protein oxidation. As measured by the introduction of carbonyl groups, protein oxidation has been associated with aging, oxidative stress and a number of diseases, such as the premature aging diseases, progeria and Werner's syndrome [34].

Protein Oxidation in the Stratum corneum

In our first methodological approach to protein oxidation in the SC, we have developed an ELISA technique to detect macromolecular carbonyls in the SC [35]. Higher carbonyl levels were detected in upper SC from environmentally exposed (tanned) skin than in nonexposed, pale skin of human volunteers. Although useful as an indicator of environmentally induced oxidative modifications in the SC, this method does not allow the differentiation between carbonyl groups derived from lipids or proteins, nor the identification of oxidized proteins. Therefore, we employed an immunoblotting technique to detect protein oxidation in human SC obtained by tape stripping.

The aim was to evaluate the relative susceptibility of human SC proteins to oxidative attack by model oxidants, to identify individual highly susceptible SC proteins and to determine physiological protein oxidation gradients within human SC.

Protein carbonyl groups were measured after lysis of SC proteins by derivatization with dinitrophenylhydrazine, separation by SDS-PAGE and immunoblotting using antibodies against dinitrophenyl groups [36]. Keratin 10, identified by use of specific antibodies and by microsequencing, was demonstrated in vitro to be oxidizable by UVA irradiation, hypochlorite and BP. In vivo, a keratin 10 oxidation gradient with low levels in the lower SC layers and about 3-fold higher contents of carbonyl groups towards the outer layers was demonstrated in forehead SC of healthy volunteers.

By analyzing the levels of protein carbonyls in human SC keratins using an immunoblotting technique, we provide direct evidence that (1) human SC keratins are susceptible to carbonyl formation by model oxidants and UVA radiation, (2) one of the oxidizable keratins was identified as keratin 10 and (3) keratin 10 oxidation increases dramatically from the inner to the outer SC in healthy forehead skin [36].

Role of Protein Oxidation Gradients in the Skin Barrier

In analogy to the high degree of saturation of fatty acids in the SC, the amount of disulfide cross-links in human SC is known to be manyfold higher than in lower epidermal layers. Similarly, we found that keratins in human SC contain dramatically more carbonyl groups than the keratins present in keratinocytes, indicating that the baseline levels of keratin oxidation are considerably higher in the SC as compared with lower epidermal layers. By using sequential tape strippings, a steep gradient with the lowest levels of carbonyl groups in keratins from lower layers and the highest in the upper layers was found. Importantly, this protein oxidation gradient is inversely correlated with the gradients of the antioxidant vitamin E [18] and free thiols [37] in human SC. There is in vitro and in vivo evidence from other biological systems that protein oxidation can be counteracted by antioxidants such as vitamin E and thiols. Furthermore, the oxygen partial pressure, a rate-limiting factor for the formation of reactive oxygen intermediates in skin [38], decreases gradually from outer to inner SC layers. Besides oxygen, the percutaneous penetration of most molecules, among them noxious, oxidizing xenobiotics, leads to a gradient within the SC with highest concentrations in the outer layers. The inverse correlation with SC antioxidant levels on the one hand and the positive correlation with the levels of oxygen and oxidizing xenobiotics within the SC on the other may account for the protein oxidation gradient in SC keratins.

Protein Oxidation and Desquamation: A Relevant Link?

We propose that the protein oxidation gradient with increased levels towards outer SC layers may have important implications for the process of desquamation. Since proteins in corneodesmosomes play a crucial role in SC cell cohesion and specific proteases have been identified in human SC, proteolysis is generally believed to be a key event in desquamation. However, since the SC consists of enucleated, 'dead' cells, it is still unclear, how the onset of desquamation in the upper SC is regulated [39]. Many common proteases degrade oxidized proteins more rapidly than unoxidized forms [32]. Thus, in addition to regulation by other factors, the higher levels of protein oxidation detected in the upper SC may account for an increased susceptibility of keratins and other macromolecules to be degraded by SC proteases, leading to desquamation in the superficial SC layers.

While protein oxidation increases proteolytic susceptibility up to a protein-specific degree, further damage actually causes a decrease in proteolytic susceptibility and leads to cross-linking and aggregation [40]. Furthermore, protein-bound carbonyl groups are believed to be involved in intra- and intermolecular cross-linking. Although it is well accepted that cross-linking of SC keratins serves to improve the physical stability of the keratin network, very little is known about the biochemical details [41]. The introduction of carbonyl groups into SC keratins is likely to have implications for keratin cross-linking, since protein cross-linking and aggregation are not limited to disulfide cross-linking, but also other forms of covalent cross-links such as the formation of an intermolecular Schiff base by reaction of a carbonyl group from one protein with an amino group from another. Possibly, transglutaminase which catalyzes the formation of an amide bond between the γ-carbonyl group of glutamine and the ε-amino-group of lysine and plays an important role in the formation of the cornified envelope, is involved in this dimerization. Notably, eye lens proteins were shown to be far more susceptible to transglutaminase-catalyzed reactions when preincubated with ROS [42].

A recent publication has provided further evidence for a link between redox properties of the SC and desquamation: An SC thiol protease, a novel cysteine protease of late epidermal differentiation, was demonstrated to be dependent on the redox potential of the SC. Due to its desquamatory action in in vitro assays and its virtual absence in palmoplantar SC, it was proposed as relevant desquamatory protease [43].

Conclusion

Our findings demonstrate that the introduction of carbonyl groups into human SC keratins is inducible by oxidants and that the levels of protein oxidation increase towards outer SC layers. These findings may contribute to a better understanding of the complex biochemical processes of keratinization and desquamation.

Microbial Colonization and Infection

The normal human skin is colonized by huge numbers of bacteria and fungi that live harmlessly as commensals on its surface and within its follicles. Most organisms reside on the surface of the SC between squames in the outermost layers. At times, overgrowth of some of these resident organisms may cause diseases of the skin.

The normal cutaneous barrier represents an effective defense mechanism against microbial infections. Predisposing factors such as mechanical injuries, abnormal humidity, immunodeficiency or metabolic disorders can lead to severe microbe-associated skin disorders. The mechanisms of cellular damage caused by infectious and inflammatory processes in skin are complex. There is, however, a consensus that ROS generated by phagocytes migrating to injured tissues might be the main agents responsible for cellular damage in inflammatory processes [44]. Intriguingly, phagocyte-generated ROS are necessary for an efficient clearance of pathogenic microorganisms.

Recent investigations have demonstrated that oxidative stress may be caused by the microbial flora of human skin. Common cutaneous pathogens, including dermatophytes and yeasts, or bacteria of the species *Streptococcus* or *Staphylococcus*, are aerobic organisms capable of generating remarkable amounts of ROS [45]. Furthermore, there is evidence that *Candida albicans*, the most important opportunistic fungal pathogen, releases hydrogen peroxide in a similar way as phagocytes. Interestingly, the generation of ROS by *C. albicans* increases significantly during formation of hyphae, which occurs mostly during their pathophysiologically relevant penetration of the skin barrier into deeper skin layers [46]. It seems likely that this mechanism involves oxidative stress in the skin barrier.

In conclusion, ROS play a central role in the pathophysiology of bacterial and fungal skin infections. Therapeutical applications of antioxidants could be very useful in the modulation of oxidative damage induced directly by microbes and the resulting oxidative burst by phagocytes.

Aging

It has been suggested that the aging process is dependent on the action of free radicals [32]. Several key metabolic enzymes are oxidatively inactivated by a variety of mixed-function oxidation. Many of the enzymes which are inactivated have been shown to accumulate as inactive or less active forms during cellular aging. Oliver et al. [34] have demonstrated that in cultured fibroblasts from normal donors the levels of oxidatively modified proteins increase only after the age of 60 years. However, the levels of oxidatively modified proteins in fibroblasts from individuals with progeria or Werner's syndrome were significantly higher than those from age-matched controls. Moreover, treatment of glucose-6-phosphate dehydrogenase with a mixed-function oxidation system was shown to lead to oxidative modification and increased heat lability of the enzyme. Taken together these results suggest that loss of functional enzyme activity and increased heat lability of enzymes during aging may be due in part to oxidative modification by mixed-function oxidation systems. Another recent study by Merker et al. [47] demonstrated that old fibroblasts are much more vulnerable to the accumulation of oxidized proteins after oxidative stress and are not able to remove these oxidized proteins as efficiently as young fibroblasts.

Synopsis

Many studies have demonstrated beneficial health effects of topical antioxidant application; however, the underlying mechanisms are not well understood. To better understand the protective mechanism of exogenous anti- oxidants, it is important to clarify the physiological distribution, activity and regulation of antioxidants. Also, the generation of ROS by the resident and transient microbial flora and their interaction with cutaneous antioxidants appears to be of relevance for the redox properties of skin.

Our studies have demonstrated that α-tocopherol is, relative to the respective levels in the epidermis, the major antioxidant in the human SC, that α-tocopherol depletion is a very early and sensitive biomarker of environmentally induced oxidation and that a physiological mechanism exists to transport α-tocopherol to the skin surface via sebaceous gland secretion. Furthermore, there is conclusive evidence that the introduction of carbonyl groups into human SC keratins is inducible by oxidants and that the levels of protein oxidation increase towards outer SC layers. The demonstration of specific redox gradients within the human SC may contribute to a better understanding of the complex biochemical processes of keratinization and desquamation.

Taken together, the presented data suggest that, under conditions of environmentally challenged skin or during prooxidative dermatological treatment, topical and/or systemic application of antioxidants could support physiological mechanisms to maintain or restore a healthy skin barrier. Growing experimental evidence should lead to the development of more powerful pharmaceutical and cosmetic strategies involving antioxidant formulations to prevent UV-induced carcinogenesis and photoaging as well as to modulate desquamatory skin disorders.

Acknowledgements

This work was supported by the Deutsche Forschungsgemeinschaft (Th 620/2-1) and by a gift from Colgate-Palmolive Company.

References

1 Thiele JJ, Podda M, Packer L: Tropospheric ozone: An emerging environmental stress to skin. Biol Chem 1997;378:1299–1305.
2 Cross CE, van der Vliet A, Louie S, Thiele JJ, Halliwell B: Oxidative stress and antioxidants at biosurfaces: Plants, skin and respiratory tract surfaces. Environ Health Perspect 1998;106(suppl 5): 1241–1251.
3 Thiele JJ, Dreher F, Packer L: Antioxidant defense systems in skin; in Elsner P, Maibach H (eds): Drugs vs Cosmetics: Cosmeceuticals? New York, Dekker, 2000, pp 145–188.
4 Elias PM: Epidermal lipids, barrier function, and desquamation. J Invest Dermatol 1983;80:44–49.
5 Rawlings AV, Scott IR, Harding CR, Bowser PA: Stratum corneum moisturization at the molecular level. J Invest Dermatol 1994;103:731–740.
6 Traber MG, Sies H: Vitamin E in humans – Demand and delivery. Annu Rev Nutr 1996;16:321–347.
7 Shindo Y, Witt E, Han D, Epstein W, Packer L: Enzymic and non-enzymic antioxidants in epidermis and dermis of human skin. J Invest Dermatol 1994;102:122–124.
8 Bommannan D, Potts RO, Guy RH: Examination of stratum corneum barrier function in vivo by infrared spectroscopy. J Invest Dermatol 1990;95:403–408.
9 Thiele JJ, Rallis M, Izquierdo-Pullido M, Traber MG, Weber S, Maibach H, Packer L: Benzoyl peroxide depletes human stratum corneum antioxidants. J Invest Dermatol 1998;110:675A.
10 Weber SU, Thiele JJ, Cross CE, Packer L: Vitamin C, uric acid and glutathione gradients in murine stratum corneum and their susceptibility to ozone exposure. J Invest Dermatol 1999;113:1128–1132.
11 Guarrera M, Ferrari P, Rebora A: Catalase in the stratum corneum of patients with polymorphic light eruption. Acta Derm Venereol 1998;78:335–336.
12 Thiele JJ, Traber MG, Tsang KG, Cross CE, Packer L: In vivo exposure to ozone depletes vitamins C and E and induces lipid peroxidation in epidermal layers of murine skin. Free Radic Biol Med 1997;23:385–391.

13 Thiele JJ, Traber MG, Podda M, Tsang K, Cross CE, Packer L: Ozone depletes tocopherols and tocotrienols topically applied to murine skin. FEBS Lett 1997;401:167–170.
14 Thiele JJ, Traber MG, Polefka TG, Cross CE, Packer L: Ozone exposure depletes vitamin E and induces lipid peroxidation in murine stratum corneum. J Invest Dermatol 1997;108:753–757.
15 Weber SU, Han N, Packer L: Ozone: An emerging oxidative stressor to skin; in Elsner P, Thiele J (eds): Oxidants and Antioxidants in Cutaneous Biology: Curr Probl Dermatol. Basel, Karger, 2000, vol 29, pp 52–61.
16 Haratake A, Uchida Y, Mimura K, Elias PM, Holleran WM: Intrinsically aged epidermis displays diminished UVB-induced alterations in barrier function associated with decreased proliferation. J Invest Dermatol 1997;108:319–323.
17 Kitazawa M, Podda M, Thiele JJ, Traber MG, Iwasaki K, Sakamoto K, Packer L: Interactions between vitamin E homologues and ascorbate free radicals in murine skin homogenates irradiated with ultraviolet light. Photochem Photobiol 1997;65:355–365.
18 Thiele JJ, Traber MG, Packer L: Depletion of human stratum corneum vitamin E: An early and sensitive in vivo marker of UV-induced photooxidation. J Invest Dermatol 1998;110:756–761.
19 Shindo Y, Witt E, Han D, Packer L: Dose-response effects of acute ultraviolet irradiation on antioxidants and molecular markers of oxidation in murine epidermis and dermis. J Invest Dermatol 1994;102:470–475.
20 Shindo Y, Witt E, Packer L: Antioxidant defense mechanisms in murine epidermis and dermis and their responses to ultraviolet light. J Invest Dermatol 1993;100:260–265.
21 Weber C, Podda M, Rallis M, Thiele JJ, Traber MG, Packer L: Efficacy of topically applied tocopherols and tocotrienols in protection of murine skin from oxidative damage induced by UV-irradiation. Free Radic Biol Med 1997;22:761–769.
22 Stoyanovsky DA, Osipov AN, Quinn PJ, Kagan VE: Ubiquinone-dependent recycling of vitamin E radicals by superoxide. Arch Biochem Biophys 1995;323:343–351.
23 Lang JK, Gohil K, Packer L: Simultaneous determination of tocopherols, ubiquinols, and ubiquinones in blood, plasma, tissue homogenates, and subcellular fractions. Anal Biochem 1986;157:106–116.
24 Blanc D, Saint-Leger D, Brandt J: An original procedure for quantification of cutaneous resorption of sebum. Arch Dermatol Res 1989;281:346–350.
25 Faergemann J, Godleski J, Laufen H, Liss RH: Intracutaneous transport of orally administered fluconazole to the stratum corneum. Acta Derm Venereol 1995;75:361–363.
26 Fuchs J: Oxidative Injury in Dermatopathology. Berlin, Springer, 1992.
27 Picardo M, Zompetta C, De Luca C, Amantea A, Faggioni A, Nazzaro-Porro M, Passi S: Squalene peroxides may contribute to ultraviolet light-induced immunological effects. Photodermatol Photoimmunol Photomed 1991;8:105–110.
28 Thiele JJ, Weber SU, Packer L: Sebaceous gland secretion is a major physiological route of vitamin E delivery to skin. J Invest Dermatol 1999;113:1006–1010.
29 Gniadecka M, Wulf HC, Mortensen NN, Poulsen T: Photoprotection in vitiligo and normal skin: A quantititative assessment of the role of stratum corneum, viable epidermis and pigmentation. Acta Derm Venereol 1996;76:429–432.
30 Pelle E, Muizzuddin N, Mammone T, Marenus K, Maes D: Protection against endogenous and UVB-induced oxidative damage in stratum corneum lipids by an antioxidant-containing cosmetic formulation. Photodermatol Photoimmunol Photomed 1999;15:115–119.
31 Black HS, Mathews-Roth MM: Protective role of butylated hydroxytoluene and certain carotenoids in photocarcinogenesis. Photochem Photobiol 1991;53:707–716.
32 Stadtman ER: Protein oxidation and aging. Science 1992;257:1220–1224.
33 Berlett BS, Stadtman ER: Protein oxidation in aging, disease, and oxidative stress. J Biol Chem 1997;272:20313–20316.
34 Oliver CN, Ahn B-W, Moerman EJ, Goldstein S, Stadtman ER: Age-related changes in oxidized proteins. J Biol Chem 1987;262:5488.
35 Thiele JJ, Traber MG, Re R, Espuno N, Yan L-J, Cross CE, Packer L: Macromolecular carbonyls in human stratum corneum: A biomarker for environmental oxidant exposure? FEBS Lett 1998;422:403–406.

36 Thiele JJ, Hsieh SN, Briviba K, Sies H: Protein oxidation in human stratum corneum: Susceptibility of keratins to oxidation in vitro and presence of a keratin oxidation gradient in vivo. J Invest Dermatol 1999;113:335–339.

37 Broekaert D, Cooreman K, Coucke P, Nsabumukunzi S, Reyniers P, Kluyskens P, Gillis E: A quantitative histochemical study of sulphydryl and disulphide content during normal epidermal keratinization. Histochem J 1982;14:573–584.

38 Fuchs J, Thiele J: The role of oxygen in cutaneous photodynamic therapy. Free Radic Biol Med 1998;24:835–847.

39 Egelrud T, Lundstroem A, Sondell B: Stratum corneum cell cohesion and desquamation in maintenance of the skin barrier; in Marzulli FN, Maibach HI (eds): Dermatotoxicology. Washington, Taylor & Francis, 1996, pp 19–27.

40 Grune T, Reinheckel T, Davies KJ: Degradation of oxidized proteins in mammalian cells. FASEB J 1997;11:526–534.

41 Pang Y-YS, Schermer A, Yu J, Sun T-T: Suprabasal change and subsequent formation of disulfide-stabilized homo- and heterodimers of keratins during esophageal epithelial differentiation. J Cell Sci 1993;104:727–740.

42 Brossa O, Seccia M, Gravela E: Increased susceptibility to transglutaminase of eye lens proteins exposed to activated oxygen species produced in the glucose-glucose oxidase reaction. Free Radic Res Commun 1990;11:223–229.

43 Watkinson A: Stratum corneum thiol protease (SCTP): A novel cysteine protease of late epidermal differentiation. Arch Dermatol Res 1999;291:260–268.

44 Ginsburg I: Could synergistic interactions among reactive oxygen species, proteinases, membrane-perforating enzymes, hydrolases, microbial hemolysins and cytokines be the main cause of tissue damage in infectious and inflammatory conditions? Med Hypotheses 1998;51:337–346.

45 Miller RA, Britigan BE: Role of oxidants in microbial pathophysiology. Clin Microbiol Rev 1997; 10:1–18.

46 Schroeter C, Hipler UC, Wilmer A, Kuenkel W, Wollina U: Generation of reactive oxygen species by *Candida albicans* in relation to morphogenesis. Arch Dermatol Res 2000;292:260–264.

47 Merker K, Sitte N, Grune T: Hydrogen peroxide-mediated protein oxidation in young and old human MRC-5 fibroblasts. Arch Biochem Biophys 2000;375:50–54.

Jens J. Thiele, MD, Department of Dermatology and Allergology,
Friedrich Schiller University, Jena, Erfurter Strasse 35, D–07740 Jena (Germany)
Tel. +49 3641 937381, Fax +49 3641 937315, E-Mail thiele@derma.uni-jena.de

Thiele J, Elsner P (eds): Oxidants and Antioxidants in Cutaneous Biology.
Curr Probl Dermatol. Basel, Karger, 2001, vol 29, pp 43–51

Activity of Alpha-Lipoic Acid in the Protection against Oxidative Stress in Skin

Maurizio Podda[a], *Thomas M. Zollner*[a], *Marcella Grundmann-Kollmann*[a], *Jens J. Thiele*[b], *Lester Packer*[c], *Roland Kaufmann*[a]

[a] Department of Dermatology, J.W. Goethe University, Frankfurt, and
[b] Department of Dermatology, F. Schiller University, Jena, Germany;
[c] Department of Molecular and Cell Biology, University of California, Berkeley, Calif., USA

Skin is constantly exposed to prooxidant environmental stresses from a wide array of sources, ranging from UV irradiation to cigarette smoke or ozone. Reactive oxygen species (ROS) have been implicated in the etiology of several skin disorders including skin cancer [1, 2] and photoaging [3, 4]. These ROS are capable of oxidizing lipids, proteins or DNA leading to the formation of oxidized products such as lipid hydroperoxides, protein carbonyls or 8-hydroxyguanosine [5–7]. Antioxidant enzymes and nonenzymatic antioxidants also termed 'small-molecular-weight antioxidants' protect cells against oxidative damage. However, if the antioxidant complement of skin is overwhelmed by the presence of ROS, it can lead to oxidative damage of cell constituents. This has been shown for UV light [5, 8, 9] and ozone [10]. An increase in cellular antioxidants can be achieved by exogenous administration of antioxidant compounds. In the case of the skin, the most appropriate route of administration seems to be the topical application, because it allows (1) to reach higher tissue levels with (2) high tissue specificity and (3) diminishes potential side effects to other organs. This, however, can only be achieved if this antioxidant penetrates the skin, is present in its active form in the skin and effectively protects there against oxidative damage.

α-Lipoic-acid (1,2-dithiolane-3-pentanoic acid or 6,8-thioctic acid, fig. 1) has been shown to be a potent antioxidant under various conditions in vitro and in vivo [reviewed in 11]. α-Lipoic acid is a naturally occurring compound, which is an essential component of oxidative metabolism, participating as

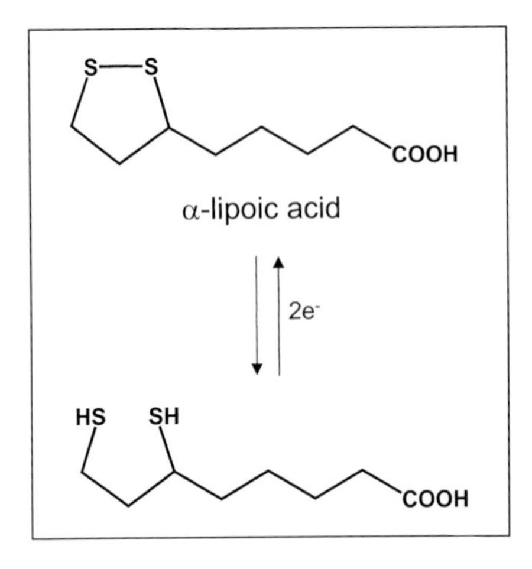

Fig. 1. Chemical structures of the redox couple α-lipoic acid and dihydrolipoic acid.

protein-bound lipoamide, in α-keto acid dehydrogenase complexes in mitochondria.

Our attention, however, focuses on the effects of the free, non-protein-bound α-lipoic acid, which can be supplied exogenously to cells or tissues. Dietary supplementation with α-lipoic acid leads to increased tissue levels of the free form. In tissues, α-lipoic acid is then reduced to dihydrolipoic acid (DHLA, fig. 1), the more active antioxidant molecule, which is partially released into the extracellular space [12].

The redox potential of the dihydrolipoate/lipoate couple is –0.32 V. For comparison, the redox potential of the glutathione/glutathione disulfide couple is –0.24 V. This means that α-lipoate/dihydrolipoate will reduce glutathione disulfide to glutathione. DHLA has been proposed to either prevent lipid peroxidation by reducing glutathione which in turn recycles vitamin E [13] or by reducing ascorbate which then reduces vitamin E [14].

α-Lipoic acid is soluble in both aqueous and lipid environments, allowing it to interact with oxidants or other antioxidants in many cellular compartments. This combination of traits makes α-lipoic acid an intriguing antioxidant for the prevention of oxidative damage to skin.

Materials and Methods

Cell Culture

Normal human keratinocytes (NHK) were either purchased (Cascade Biologics Inc., Portland, OR, USA) or were isolated from adult human skin and cultured in serum-free keratinocyte medium (Gibco, Karlsruhe, Germany). NHK were incubated with different nontoxic concentrations of α-lipoic acid (1–10 m*M*). α-Lipoic acid was a kind gift of Asta Medica (Frankfurt am Main, Germany). Irradiated cells were incubated with α-lipoic acid for 1 h prior to UV irradiation. Cells were irradiated with 25 J/cm^2 UVA light, harvested and extracted directly after UV irradiation. The UVA irradiation equipment consisted of a UVA lamp (Waldmann, Villingen-Schwenningen, Germany) emitting exclusively UVA in the range of 320–400 nm with a peak at 365 nm.

α-Lipoic Acid and Dihydrolipoic Acid Determination

α-Lipoic acid and DHLA were extracted and measured as previously described [12]. Briefly, cells or tissues were homogenized for 1 min in 1 ml of 3.3% sulfosalicylic acid and 5 m*M* EDTA. Then 1 ml of ethanol was added and thoroughly mixed for 1 min and centrifuged at 3,000 g for 3 min. The supernatant was directly injected into the HPLC column and determined using a dual gold/mercury electrode with electrochemical detection, as described by Han et al. [15].

Determination of Lipid-Soluble Antioxidants

Cells were extracted and analyzed for tocopherol and ubiquinol/ubiquinone content as previously described [16]. Briefly, cells were homogenized in 1 ml PBS and extracted into ethanol/hexane, dried down under nitrogen and after resuspension in ethanol/methanol 1:1 injected into the HPLC column. Tocopherol and ubiquinol were determined by electrochemical detection, ubiquinone by in-line UV detection.

Electrophoretic Mobility Shift Assays

Total cell extracts from 1×10^7 NHK were prepared by resuspending PBS-washed cell pellets with a buffer containing the detergent Nonidet P-40/Igepal CA-630, as described by Korner et al. [17]. Protein concentrations of supernatants were determined according to the method of Bradford (Bio-Rad, München, Germany). DNA binding conditions for nuclear factor (NF) κB have been described in detail previously [18]. Briefly, 10 μg of protein were used for the binding reaction, which contained 10,000 cpm of ^{32}P-labeled, double-stranded oligonucleotide with a high-affinity NF-κB binding motif from the κ light chain enhancer (Promega, Mannheim, Germany). DNA binding reactions were analyzed by electrophoresis on native 4% polyacrylamide gels.

Results and Discussion

Conversion of α-Lipoic Acid to Dihydrolipoic Acid by Normal Human Keratinocytes and Murine Skin

NHK were incubated in the presence of 0.5, 2 and 5 mM α-lipoic acid. After 2 h incubation a substantial amount of α-lipoic acid was converted to DHLA (fig. 2a). Furthermore, DHLA was released into the medium (fig. 2b). Handelman et al. [19] showed that leukocytes and fibroblasts are capable of taking up α-lipoic acid and partially convert it to DHLA. Haramaki et al. [20] reported that the mechanisms of reduction of α-lipoic acid are highly

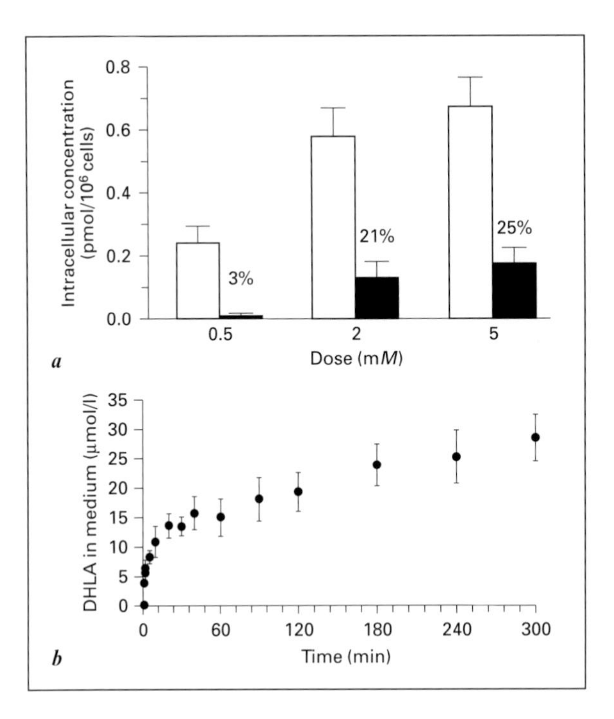

Fig. 2. ***a*** Dose-dependent increase in intracellular concentration of α-lipoic acid (0.5–5 m*M*) and DHLA in human keratinocytes. White bars indicate α-lipoic acid and black bars DHLA. Figures give the percentage of α-lipoic acid converted to intracellular DHLA (means $\pm$ SD, n = 3). ***b*** Release of DHLA into cell culture medium in the presence of keratinocytes. 2 m*M* α-lipoic acid was added to the culture medium. Already after 30 s, small amounts of DHLA are detectable in the medium (means $\pm$ SD, n = 3).

tissue specific and effects of exogenously supplied α-lipoic acid are determined by tissue glutathione reductase and dihydrolipoamide dehydrogenase activity. In the case of keratinocytes it is of further interest whether this uptake and reduction would also be relevant in vivo. Indeed, we have shown that murine skin is capable of reducing α-lipoic acid to DHLA. Two hours after application of 5% α-lipoic acid in propylene glycol (240 m*M*) we found 300 $\pm$ 122 nmol/g skin with 5% (15 $\pm$ 8 nmol/g skin) converted to DHLA [21]. Since α-lipoic acid penetrates rapidly into murine and human skin down to the dermal layers, it is possible that both keratinocytes and fibroblasts are responsible for the reduction to DHLA. Based on these findings it seems possible to assume that α-lipoic acid could be a good candidate for topical application and protection against oxidative stress in skin.

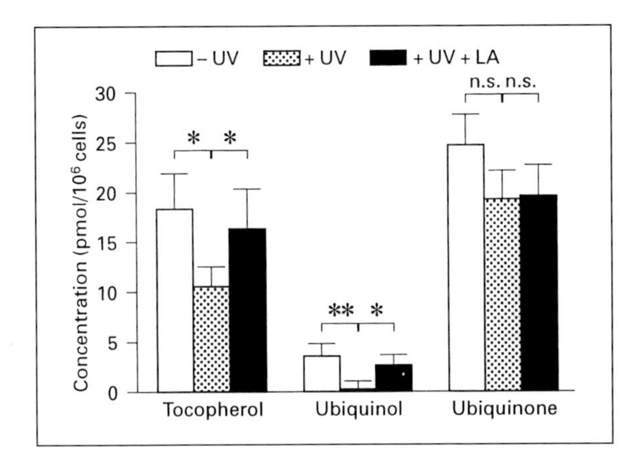

Fig. 3. Decrease in lipid-soluble antioxidants (tocopherol, ubiquinol and ubiquinone) in keratinocytes after UVA irradiation (25 J/cm^2). 2 m*M* α-lipoic acid (LA) inhibits the loss of tocopherol and ubiquinol (means ± SD, n = 4, $^{**}p < 0.01$, $^{*}p < 0.5$).

Protection against Loss of Lipid-Soluble Antioxidants in Normal Human Keratinocytes Treated with UVA Light

NHK were treated with increasing doses of UVA light. At 25 J/cm^2 we could observe a decrease in both the ubiquinol/ubiquinone and tocopherol levels in NHK. Pretreatment with 0.5–5 m*M* α-lipoic acid led to a dose-dependent protection against the decrease in ubiquinol/ubiquinone and tocopherol in NHK (fig. 3).

It has previously been shown that UV light is capable of exerting an oxidative stress [22] and in relevant doses overwhelms the antioxidant network defense system present in the skin [8, 9]. One of the early events is the decrease in lipid-soluble antioxidants as we have shown for human skin equivalents irradiated with light from a solar simulator [5]. In this experiment we have used exclusively UVA light. Under this condition, α-lipoic acid was highly effective in the protection against antioxidant loss. There is precedence for this type of observation as we had previously shown that α-lipoic acid was capable of protecting against antioxidant loss in murine skin. Both tocopherol and ubiquinol were reduced after UV irradiation, and α-lipoic acid significantly protected against the loss of ubiquinol [23]. Interestingly, both in NHK and in the murine in vivo experiments a strong decrease in α-lipoic acid levels could be observed after UV irradiation suggesting that it is partially used up while protecting other antioxidants against irreversible oxidation. However, DHLA levels did not differ to such a large extent before and after UV irradiation. This could be either due to a constant and rapid replacement of DHLA from the α-lipoic acid pool or possibly because UV

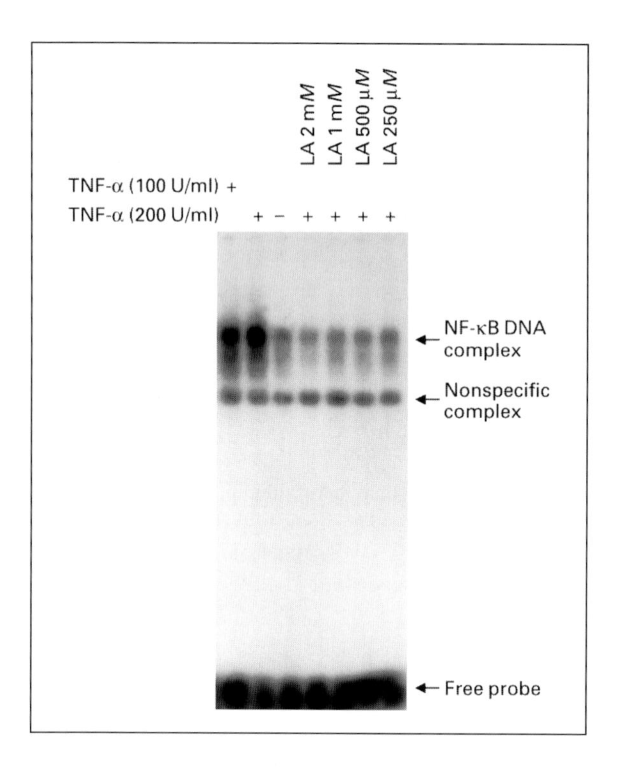

Fig. 4. α-Lipoic acid (LA) inhibits in a dose-dependent manner TNF-α-induced NF-κB activation in human keratinocytes. Representative electromobility shift assay of 3 independent experiments.

irradiation produces preferentially ROS which are better scavenged by α-lipoic acid than by DHLA. A third possibility is that α-lipoic acid is reduced to DHLA by a photochemical reaction. This, however, seems excluded since we did not find significant amounts of DHLA after irradiation of cell culture medium with α-lipoic acid without cells.

α-Lipoic Acid Inhibits the Activation of the Redox-Sensitive Transcription Factor NF-κB

To further evaluate the actual functional relevance of the antioxidant activity of α-lipoic acid in NHK we determined the efficacy of α-lipoic acid in reducing the activation of the transcription factor NF-κB. NF-κB is a protein transcription factor that is required for maximal transcription of a wide array of proinflammatory molecules [24]. It consists of a heterodimer of the p50 and p65 proteins retained inactive in the cytoplasm tightly bound to the inhibitory subunit IκB-α. Upon activation, IκB-α is rapidly and sequentially phosphorylated by the action of IκB kinases, ubiquitinated and degraded by proteasomes [25]. The active subunit is translocated to the nucleus where it binds to cognate DNA sequences. It is responsible for the regulation of a vast array of proinflammatory genes. Antioxidants, like N-acetyl-L-cysteine, are believed to interfere with the most proximal step in NF-κB activation. They inhibit NF-κB activation via suppression of IκB kinase activity preventing IκB phosphorylation and degradation [26]. An inhibition of NF-κB activation by α-lipoic acid has previously been shown in the Jurkat lymphocyte cell line [27] and in the HaCaT keratinocyte cell line [28]. We found a strong and dose-dependent inhibition of the activation of the transcription factor NF-κB in NHK (fig. 4). This will presumably lead to a decreased expression of NF-κB-regulated genes, like ICAM-1, TNF-α or IL-6. It is known from other antioxidant compounds, that they indeed inhibit proinflammatory cytokines or adhesion molecules via NF-κB [29–32]. For α-lipoic acid, however, this still needs to be shown.

Conclusion

α-Lipoic acid is a powerful antioxidant, which directly scavenges a large array of ROS. Furthermore, α-lipoic acid is capable of interacting with both water- and lipid-soluble antioxidants and indirectly bolstering the antioxidant defense network. In the case of skin, keratinocytes and fibroblasts reduce α-lipoic acid to DHLA which is an even more active antioxidant. Since α-lipoic acid easily and rapidly enters murine and human skin, it can be assumed that it is present in amounts needed for effective antioxidant protection. Indeed, we observed that 2 m*M* α-lipoic acid significantly protect human keratinocytes against UV-induced oxidative damage. This antioxidant protection was of actual functional relevance as shown by the inhibition of NF-κB activation. α-Lipoic acid is not the only antioxidant with a good efficacy in several model systems. Tocopherol, ascorbate and other natural or synthetic antioxidants have been shown to exert an antioxidant effect in skin or skin cells [33–37]. However, based on our in vitro and in vivo studies we suggest that α-lipoic acid could be a good candidate antioxidant for the protection of skin against oxidative damage. Further experiments are warranted to determine the extent of actual effects against photoaging and ROS-mediated carcinogenesis in skin.

Acknowledgments

We thank Beth Koh and Derick Han for helpful assistance in setting up the α-lipoic acid HPLC analysis. Part of this work was supported by ASTA Medica, USA, and the Marie Christine Held and Erika Hecker Foundation (M.P.).

References

1 Guyton KZ, Kensler TW: Oxidative mechanisms in carcinogenesis. Br Med Bull 1993;49:523–544.
2 Perchellet JP, Perchellet EM: Antioxidants and multistage carcinogenesis in mouse skin. Free Radic Biol Med 1989;7:377–408.
3 Dalle Carbonare M, Pathak MA: Skin photosensitizing agents and the role of reactive oxygen species in photoaging. J Photochem Photobiol B 1992;14:105–124.
4 Emerit I: Free radicals and aging of the skin. EXS 1992;62:328–341.
5 Podda M, Traber MG, Weber C, Yan L-J, Packer L: UV-irradiation depletes antioxidants and causes oxidative damage in a model of human skin. Free Radic Biol Med 1998;24:55–65.
6 Hu ML, Tappel AL: Potentiation of oxidative damage to proteins by ultraviolet-A and protection by antioxidants. Photochem Photobiol 1992;56:357–363.
7 Beehler BC, Przybyszewski J, Box HB, Kulesz-Martin MF: Formation of 8-hydroxydeoxyguanosine within DNA of mouse keratinocytes exposed in culture to UVB and H_2O_2. Carcinogenesis 1992; 13:2003–2007.
8 Shindo Y, Witt E, Packer L: Antioxidant defense mechanisms in murine epidermis and dermis and their responses to ultraviolet light. J Invest Dermatol 1993;100:260–265.
9 Fuchs J, Huflejt ME, Rothfuss LM, Wilson DS, Carcamo G, Packer L: Impairment of enzymic and nonenzymic antioxidants in skin by UVB irradiation. J Invest Dermatol 1989;93:769–773.
10 Thiele JJ, Traber MG, Tsang K, Cross CE, Packer L: In vivo exposure to ozone depletes vitamins C and E and induces lipid peroxidation in epidermal layers of murine skin. Free Radic Biol Med 1997;23:385–391.
11 Packer L, Witt EH, Tritschler HJ: Alpha-lipoic acid as a biological antioxidant. Free Radic Biol Med 1995;19:227–250.
12 Podda M, Tritschler HJ, Ulrich H, Packer L: Alpha-lipoic acid supplementation prevents symptoms of vitamin E deficiency. Biochem Biophys Res Commun 1994;204:98–104.
13 Bast A, Haenen GR: Interplay between lipoic acid and glutathione in the protection against microsomal lipid peroxidation. Biochim Biophys Acta 1988;963:558–561.
14 Kagan VE, Shvedova A, Serbinova E, Khan S, Swanson C, Powell R, Packer L: Dihydrolipoic acid – A universal antioxidant both in the membrane and in the aqueous phase: Reduction of peroxyl, ascorbyl and chromanoxyl radicals. Biochem Pharmacol 1992;44:1637–1649.
15 Han D, Handelman GJ, Packer L: Analysis of reduced and oxidized lipoic acid in biological samples by high-performance liquid chromatography. Methods Enzymol 1995;251:315–325.
16 Podda M, Weber C, Traber M, Packer L: Simultaneous determination of tissue tocopherols, tocotrienols, ubiquinols, and ubiquinones. J Lipid Res 1996;37:893–901.
17 Korner M, Rattner A, Mauxion F, Sen R, Citri Y: A brain-specific transcription activator. Neuron 1989;3:563–572.
18 Schmidt KN, Podda M, Packer L, Baeuerle PA: Anti-psoriatic drug anthralin activates transcription factor NF-kappa B in murine keratinocytes. J Immunol 1996;156:4514–4519.
19 Handelman GJ, Han D, Tritschler H, Packer L: Alpha-lipoic acid reduction by mammalian cells to the dithiol form, and release into the culture medium. Biochem Pharmacol 1994;47:1725–1730.
20 Haramaki N, Han D, Handelman GJ, Tritschler HJ, Packer L: Cytosolic and mitochondrial systems for NADH- and NADPH-dependent reduction of alpha-lipoic acid. Free Radic Biol Med 1997; 22:535–542.
21 Podda M, Rallis M, Traber MG, Packer L, Maibach HI: Kinetic study of cutaneous and subcutaneous distribution following topical application of [7,8-^{14}C]rac-alpha-lipoic acid onto hairless mice. Biochem Pharmacol 1996;52:627–633.
22 Jurkiewicz BA, Buettner GR: EPR detection of free radicals in UV-irradiated skin: Mouse versus human. Photochem Photobiol 1996;64:918–922.
23 Podda M, Traber M, Packer L: α-Lipoate: Antioxidant properties and effects on skin; in Fuchs J, Packer L, Zimmer G (eds): Lipoic Acid in Health and Disease. New York, Dekker, 1997, pp 163–180.
24 Baeuerle PA, Henkel T: Function and activation of NF-κB in the immune system. Annu Rev Immunol 1994;12:141–179.

25 Regnier CH, Song HY, Gao X, Goeddel DV, Cao Z, Rothe M: Identification and characterization of an IkappaB kinase. Cell 1997;90:373–383.
26 Spiecker M, Darius H, Kaboth K, Hubner F, Liao JK: Differential regulation of endothelial cell adhesion molecule expression by nitric oxide donors and antioxidants. J Leukoc Biol 1998;63: 732–739.
27 Suzuki YJ, Aggarwal BB, Packer L: Alpha-lipoic acid is a potent inhibitor of NF-kappa B activation in human T cells. Biochem Biophys Res Commun 1992;189:1709–1715.
28 Saliou C, Kitazawa M, McLaughlin L, Yang JP, Lodge JK, Tetsuka T, Iwasaki K, Cillard J, Okamoto T, Packer L: Antioxidants modulate acute solar ultraviolet radiation-induced NF-kappa-B activation in a human keratinocyte cell line. Free Radic Biol Med 1999;26:174–183.
29 Baeuml H, Behrends U, Peter RU, Mueller S, Kammerbauer C, Caughman SW, Degitz K: Ionizing radiation induces, via generation of reactive oxygen intermediates, intercellular adhesion molecule-1 (ICAM-1) gene transcription and NF kappa B-like binding activity in the ICAM-1 transcriptional regulatory region. Free Radic Res 1997;27:127–142.
30 Eberlein-Konig B, Jager C, Przybilla B: Ultraviolet B radiation-induced production of interleukin 1alpha and interleukin 6 in a human squamous carcinoma cell line is wavelength-dependent and can be inhibited by pharmacological agents. Br J Dermatol 1998;139:415–421.
31 Dy LC, Pei Y, Travers JB: Augmentation of ultraviolet B radiation-induced tumor necrosis factor production by the epidermal platelet-activating factor receptor. J Biol Chem 1999;274:26917–26921.
32 Lange RW, Hayden PJ, Chignell CF, Luster MI: Anthralin stimulates keratinocyte-derived proinflammatory cytokines via generation of reactive oxygen species. Inflamm Res 1998;47:174–181.
33 Jurkiewicz BA, Bissett DL, Buettner GR: Effect of topically applied tocopherol on ultraviolet radiation-mediated free radical damage in skin. J Invest Dermatol 1995;104:484–488.
34 Darr D, Combs S, Dunston S, Manning T, Pinnell S: Topical vitamin C protects porcine skin from ultraviolet radiation-induced damage. Br J Dermatol 1992;127:247–253.
35 Dreher F, Gabard B, Schwindt DA, Maibach HI: Topical melatonin in combination with vitamins E and C protects skin from ultraviolet-induced erythema: A human study in vivo. Br J Dermatol 1998;139:332–339.
36 Pelle E, Muizzuddin N, Mammone T, Marenus K, Maes D: Protection against endogenous and UVB-induced oxidative damage in stratum corneum lipids by an antioxidant-containing cosmetic formulation. Photodermatol Photoimmunol Photomed 1999;15:115–119.
37 Zhao J, Sharma Y, Agarwal R: Significant inhibition by the flavonoid antioxidant silymarin against 12-O-tetradecanoylphorbol 13-acetate-caused modulation of antioxidant and inflammatory enzymes, and cyclooxygenase 2 and interleukin-1alpha expression in SENCAR mouse epidermis: Implications in the prevention of stage I tumor promotion. Mol Carcinog 1999;26:321–333.

Dr. Maurizio Podda, Department of Dermatology, J.W. Goethe University of Frankfurt,
Theodor-Stern-Kai 7, D–60590 Frankfurt (Germany)
Tel. +49 69 6301 6845, Fax +49 69 6301 7139, E-Mail Podda@em.uni-frankfurt.de

Thiele J, Elsner P (eds): Oxidants and Antioxidants in Cutaneous Biology.
Curr Probl Dermatol. Basel, Karger, 2001, vol 29, pp 52–61

Ozone: An Emerging Oxidative Stressor to Skin

Stefan U. Weber, Nancy Han, Lester Packer

Department of Molecular and Cellular Biology, University of California, Berkeley, Calif., USA

Normally found in the stratosphere where it plays a beneficial role in filtering out the short-wave spectrum of the UV radiation, O_3 can be toxic when present at ground level as part of photochemical smog [1]. It is present at concentrations between 0.1 and 0.5 ppm and currently poses a severe urban air quality problem [2]. In addition to photochemical smog, O_3 can be generated during the operation of high-voltage devices [1]. Reacting readily with biomolecules, O_3, with its standard redox potential of +2.07 mV, is one of the most powerful oxidants known, having the ability to cause ozonation, oxidation and peroxidation of biomolecules, both directly and via secondary reactions [3].

The skin, as the outermost part of the body, is a primary target of environmental stressors [4], and O_3 is probably the most reactive chemical to which the skin is routinely exposed in the environment [5]. Protected from oxidative stress by a networking system of nonenzymatic (e.g. vitamin E, ubiquinone, vitamin C, glutathione) and enzymatic (e.g. superoxide dismutase, catalase, glutathione reductase and peroxidase) antioxidants [6], the skin acts as a barrier against oxidative air pollutants. By compromising antioxidant defenses and inducing oxidative damage to lipids and proteins, O_3 may affect the structural integrity of the stratum corneum (SC) [7].

This chapter will focus on O_3 reactions in the SC, investigating damage to biomolecules such as lipids and proteins, and will discuss potential physiological implications.

O_3 Reacts in the Stratum corneum

Since O_3 is a strong oxidant, it reacts quickly with biological targets. In the lung it has been calculated that most of the O_3 reacts with the lung lining

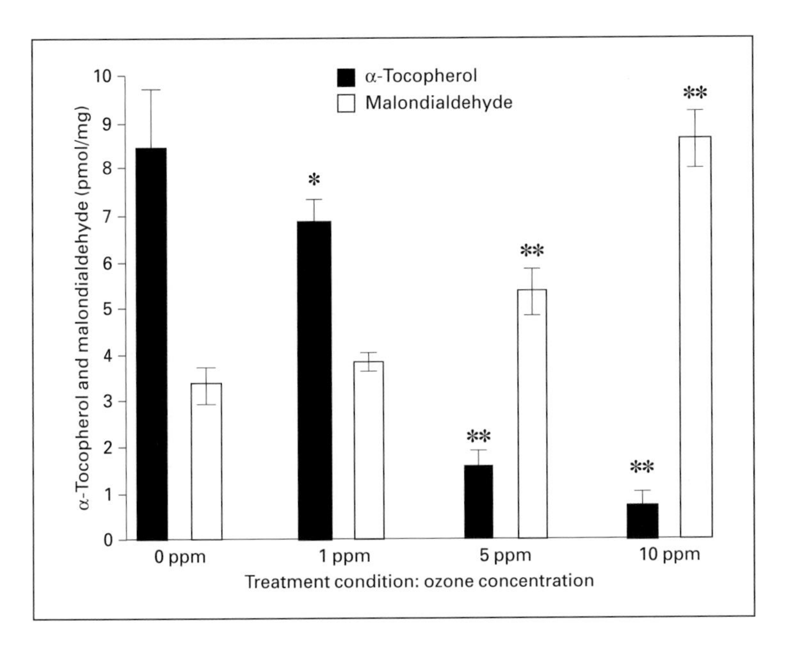

Fig. 1. Ozone depletes vitamin E and induces malondialdehyde formation in the SC in a dose-dependent manner. Mice were exposed to O_3 for 2 h (n = 12 for 0 ppm, n = 4 for each group for 1, 5 and 10 ppm). α-Tocopherol and malondialdehyde concentrations in the SC were measured by HPLC analysis. *$p < 0.05$, **$p < 0.001$. Reprinted by permission of Blackwell Science Inc. [12].

fluid and does not even reach the alveolar cells [8]. Initial studies in a hairless mouse model with high doses of O_3 (10 ppm × 2 h) were unable to detect changes in vitamin E, a major lipophilic antioxidant, when whole skin was analyzed [9]. However, topical supplementation with vitamin E in the form of tocopherols and tocotrienols resulted in a depletion of more than two thirds of the applied vitamin E [9]. In an attempt to localize the site of O_3-induced skin damage, whole skin was separated into 3 layers: upper epidermis, lower epidermis/papillary dermis and dermis. After exposure to the same dose of O_3, endogenous α-tocopherol was found to be depleted in the upper epidermis only while deeper layers remained unaffected [10]. The same pattern was observed for the hydrophilic antioxidant vitamin C. To further characterize the site of interaction, a tape-stripping method to analyze vitamin E in SC layers was developed [11]. This method allowed the detection of as little as 0.1 pmol in SC extract. It could be demonstrated that doses ranging from 10 ppm × 2 h to only 1 ppm × 2 h deplete endogenous α-tocopherol in murine SC in a dose-dependent manner [12] (fig. 1). From these studies it can be concluded that O_3 mainly reacts within the SC. These results are in accordance with findings from plant experiments where it was shown that after exposure of leaves to O_3, the O_3 concentration was zero in the intercellular space of leaves [13]. Mathematical models predict that O_3 at environmentally relevant doses reacts even within a membrane bilayer. Thus, it is not surprising that O_3 does not pass through the tightly packed SC which consists of stacks of lipid membranes and enucleated corneocytes [14]. As mentioned above, similar findings were reported for the lung, where probably all O_3 reacts with lining fluids before reaching viable airway epithelium [8, 15, 16]. In parallel, O_3 reacts in protective, nonviable skin layers without reaching viable epidermal cell layers.

O_3 Depletes Hydrophilic Antioxidants

In order to protect viable tissue from oxidative stressors such as O_3, the skin houses a variety of hydrophobic and hydrophilic antioxidants that form a defensive network [17]. Among the chief hydrophilic antioxidants are vitamin C, glutathione and uric acid [18]. Ascorbic acid appears to react directly with O_3 and is an effective O_3 scavenger in the lung, providing defense against O_3 attack and thus preventing lipid peroxidation [3]. Furthermore, vitamin C also has a strong reducing potential, thereby having the ability to regenerate α-tocopherol from the tocopheroxyl radical [19].

A tape stripping method and modified extraction using HPLC were used to detect the presence, distribution and susceptibility to O_3 exposure of these three hydrophilic antioxidants in murine SC [11] (fig. 2). The study found that the concentrations of vitamin C, GSH and urate formed a steep gradient with their content increasing towards deeper layers [7]. Even at levels as low as 1 ppm $\times$ 2 h, the susceptibility of the hydrophilic antioxidants to O_3 was evident. All three antioxidants exhibited their first significant difference in levels as compared to standard levels after exposure to 1 ppm $O_3 \times 2$ h [7]. Vitamin C was decreased to about 80% of its pretreatment level, GSH to about 40% and uric acid to about 45% in the SC after exposure to O_3, providing further evidence that O_3 depletes protective antioxidants and thus induces oxidative stress in the superficial skin layers.

O_3-Induces Damage to Biomolecules

Apart from depletion of antioxidants, O_3 has been demonstrated to damage important classes of biomolecules. Since a big portion of the SC consists

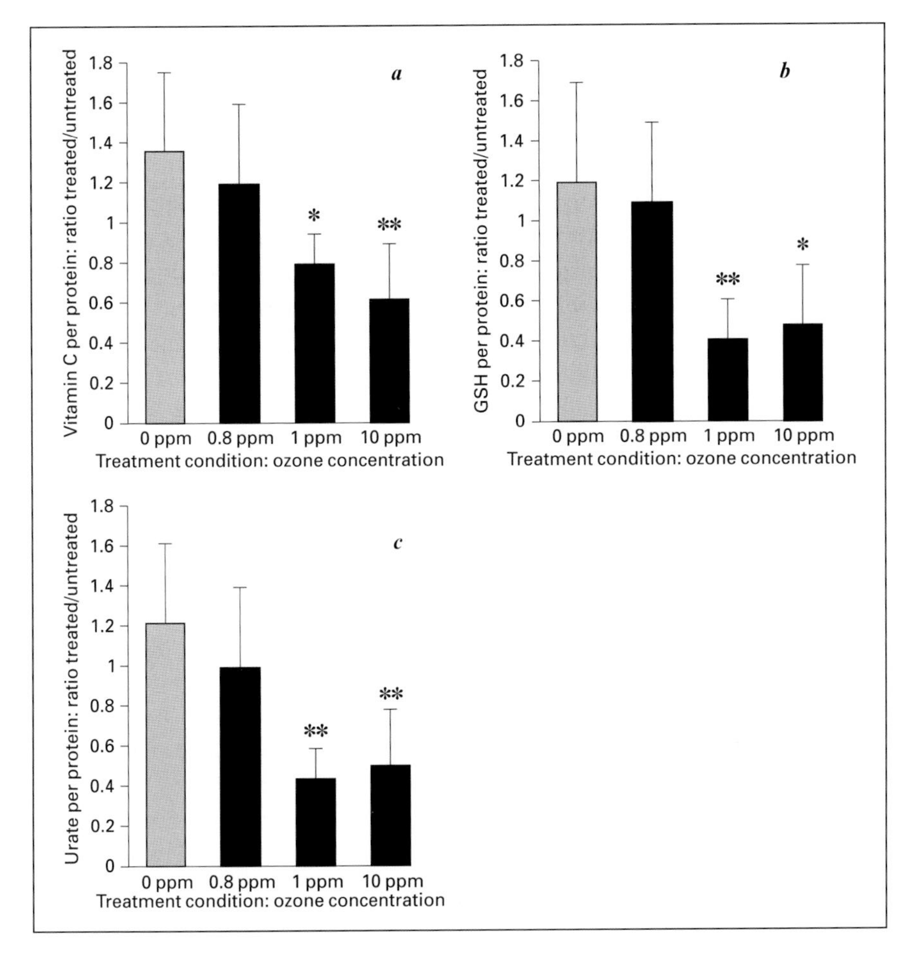

Fig. 2. Ozone depletes vitamin C(***a***), GSH (***b***) and urate (***c***) in the SC. Mice were exposed to O_3 for 2 h (n = 6) and tape stripped before and after treatment. Vitamin C, GSH and urate showed the first significant difference at 1 ppm. The graph displays the ratio of the antioxidant concentration after to that before treatment. $^*p < 0.05$, $^{**}p < 0.01$. Reprinted by permission of Blackwell Science Inc. [7].

of lipids, it was reasonable to assay for lipid peroxidation. For this purpose, an HPLC-based method to detect the thiobarbituric acid adduct of malondialdehyde (MDA) in SC tape strippings was developed [20]. MDA is a breakdown product of oxidized lipids arising during lipid peroxidation. In whole skin, 10 ppm × 2 h O_3 significantly increased the MDA content per wet weight [9]. Again, the main effect was localized in the SC.

Doses of 10 ppm × 2 h and 5 ppm × 2 h raised MDA levels significantly. However, a dose of 1 ppm × 2 h that was capable of depleting vitamin E was insufficient to increase MDA (fig. 1). This indicates that vitamin E can protect the SC from O_3-induced lipid peroxidation when relatively low doses of the gaseous oxidant are used. In contrast to findings in low-density lipoprotein particles, where a total depletion of vitamin E precedes lipid peroxidation [21], this was not the case in the skin. In the SC, a significant production of MDA was already observed when α-tocopherol was depleted to approximately 20% [12].

This discrepancy may be explained by the reaction mechanisms of O_3 with its target molecules. O_3 can react with polyunsaturated fatty acids containing at least two methylene interrupted double bonds and result in MDA generation [22]. This pathway was not inhibited by vitamin E [23]. Therefore, this reaction may occur whether or not α-tocopherol is present. Still, α-tocopherol may exert certain protective effects by inhibiting the reaction of O_3 with polyunsaturated fatty acids containing at least three double bonds, such as linolenic acid [24].

Since O_3 seems to react in the SC, the question arises if oxidized products can be found in lower layers. Using the same murine model and a dose of 10 ppm × 2 h, a massive increase in upper epidermal MDA was detected, and even in the lower epidermis/papillary dermis, a significant increase was observed [10]. These findings indicate that lipid ozonation products may travel from their origin, the SC, to deeper layers of the skin presumably by diffusion. In this way, reactive aldehydes generated in the SC may bridge the distance to viable cell layers and exert toxic effects in the epidermis.

Apart from lipids, relatively insoluble proteins account for a major portion of the SC. Corneocytes consist of core and envelope proteins which interact with the surrounding lipid layers, contributing to the mechanical stability of the skin barrier. Oxidation of amino acid residues via several mechanisms often leads to introduction of carbonyl groups. These carbonyl groups are a hallmark of protein oxidation and can be measured. SC obtained by tape stripping was exposed to O_3 in vitro and assayed for carbonyl groups. This assay measures lipid carbonyl groups as well as protein carbonyls. It could be observed that the total carbonyl content was increased by exposure to O_3 doses of 1, 5 and 10 ppm × 2 h [25]. It is reasonable to assume that a considerable portion of this increase was due to protein oxidation. To summarize, topical O_3 exposure results in lipid and most likely also in protein oxidation.

O_3 and UV Radiation Have Additive Effects

It has long been recognized that UVB, and to a much lesser extent UVA, induce a number of pathological conditions in the skin by directly oxidizing target molecules or by the formation of reactive oxygen species such as singlet oxygen. It has been documented that O_3 exposure of as little as 1 ppm $O_3 \times 2$ h depletes small-molecular-weight antioxidants in the SC [26]. While exposure of cutaneous tissue to either UV radiation or O_3 alone is enough to deplete vital antioxidants and induce lipid peroxidation, there is evidence that parallel exposure to both O_3 and solar UVA and UVB radiation can have oxidative effects on the skin.

To investigate the additive effects of both environmental oxidative stressors on the SC, a recent study was carried out by exposing hairless mice to UV radiation and O_3, alone and in combination. This study used a tape stripping method to analyze vitamin E. An O_3 dose of 1 ppm for 2 h depleted SC vitamin E to almost 40% of the values in untreated skin while 0.5 MED of UV significantly decreased the amount of α-tocopherol, but a combination of the two did not increase the effect of UV alone. However, a pretreatment of 0.5 ppm $O_3 \times 2$ h, which had no effect when used alone, followed by low-dose UV radiation (0.33 MED) significantly enhanced the UV-induced depletion of vitamin E [26]. Therefore, the additive effect of UV radiation and O_3 is most prominent with lower doses of UV radiation (fig. 3).

Whereas vitamin E loss was observed at lower levels of solar-simulated UV, an increase in MDA, a by-product of lipid peroxidation, was observed only when at least 0.5 MED solar-simulated UV was used and a combination of the two oxidant stressors did not increase the MDA content [26], adding further evidence to the tentative role of vitamin E as a sacrificial and chain-breaking antioxidant in the prevention of progressive lipid peroxidation in cutaneous tissue.

Ultimately, it can be concluded that concurrent exposure to O_3 and UV radiation at concentration levels near those that humans may be subject to in areas polluted with photochemical smog, can exhibit additive effects in terms of oxidative damage to the skin barrier.

Potential Pathophysiological Consequences of Topical O_3 Exposure

It is clear that O_3 oxidizes targets in the SC. Do these changes translate into pathophysiological events in viable epidermal cell layers? To date this question has not been fully answered.

The SC plays an important role as a diffusion barrier against exogenous chemicals and endogenous water. Disruption of the skin barrier is known to increase the transepidermal water loss. A perturbed skin barrier has been implicated in skin pathologies such as psoriasis, atopic dermatitis, irritant contact dermatitis and aging skin including pruritus of the aging skin

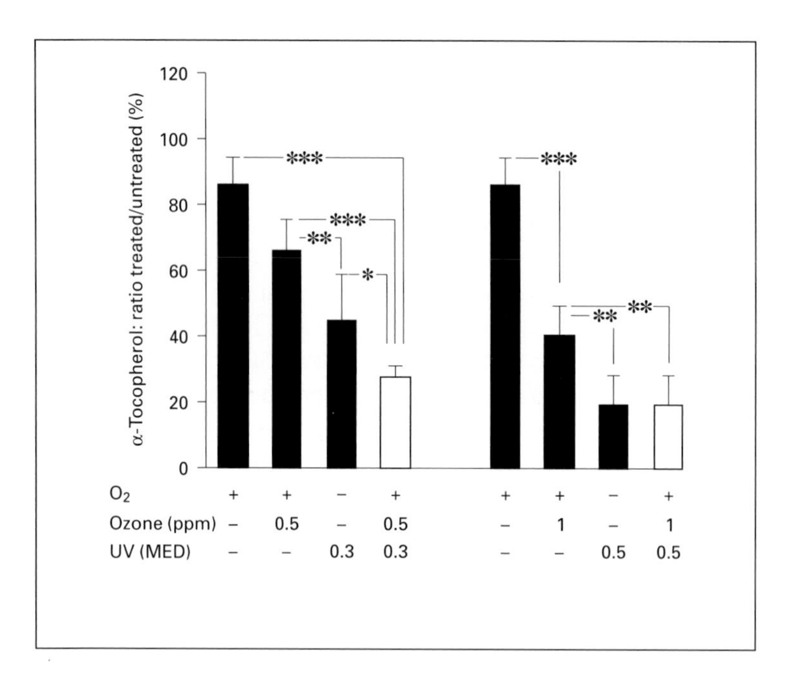

Fig. 3. Ozone and UV radiation exert additive effects in the SC. Mice were exposed to UV and O_3 and tape stripped. α-Tocopherol concentrations were analyzed by HPLC. The additive effect of combined stressors was significant with low concentrations of UV (0.3 MED) and O_3 (0.5 ppm × 2 h) when compared to either UV treatment or O_3 treatment alone. $^{*}p<0.05$, $^{**}p<0.01$, $^{***}p<0.001$. Reprinted by permission of Elsevier Science [26].

[14, 27, 28]. The composition of the SC lipid phase has been shown to be a critical determinant of the barrier function. A lack of free fatty acids, cholesterol or ceramides results in an increased transepidermal water loss [27]. O_3 is known to oxidize SC lipids. This may lead to an imbalance of critical SC lipids which in turn could translate into a disturbed skin barrier.

A perturbed skin barrier is also known to evoke repair and proinflammatory responses. Epidermal DNA synthesis is upregulated [29], and lamellar bodies are secreted from the upper epidermal cells to restore the skin barrier [30]. Moreover, barrier disruption also induces epidermal production of the cytokines IL-1α, TNF and GM-CSF, while occlusion directly after barrier disruption prevents this proinflammatory response [31–33]. If O_3 compromises the skin barrier, it may also stimulate epidermal cytokine production.

It was demonstrated that topical O_3 exposure results in formation of MDA [10]. Presumably other reactive aldehydes are also formed. In model reactions of lung lining fluid, O_3 treatment resulted in the formation of several reactive aldehydes [15]. Their presence was also confirmed in actual human lung lining fluid after O_3 exposure [34]. These aldehydes arising from ozonation cascades are believed to be the mediators of O_3 toxicity in the lung [34]. In analogy to these findings, we postulate that a similar phenomenon may occur as well in the skin. Lipid ozonation products from the SC may reach viable epidermal cells and exert biological effects in keratinocytes. In the lung it was recently shown that O_3 activates the transcription factor NF-κB [35]. Similar responses may occur in the skin.

Conclusion

There is mounting evidence that tropospheric O_3 exposure not only plays a role in lung toxicity but also affects cutaneous tissues. Since O_3 reacts readily with the closest targets, it exerts its main oxidizing effect in the outermost layer of the skin, the SC. O_3 has been shown to deplete the low-molecular-weight antioxidants α-tocopherol, vitamin C, glutathione and uric acid dose dependently. Moreover, MDA levels are elevated after O_3 exposure as an indicator of lipid peroxidation. Lipid peroxidation products were not only found in the SC but also in the neighboring viable epidermis. A combination of O_3 and UV radiation has been demonstrated to deplete SC α-tocopherol in an additive fashion when low doses were applied. These doses of 0.5 ppm $O_3 \times 2$ h can be achieved in real life situations in heavily polluted areas. However, at this point it is not known if O_3 at environmentally relevant levels exerts any effects other than oxidation of SC molecules. Possibly, O_3 may compromise the skin barrier function and may induce an epidermal proinflammatory response. Further human studies will be needed to elucidate the pathophysiological effects of O_3 on cutaneous tissues.

References

1 Mustafa MG: Biochemical basis of ozone toxicity. Free Radic Biol Med 1990;9:245–265.
2 Cross CE, van der Vliet A, Louie S, Thiele JJ, Halliwell B: Oxidative stress and antioxidants at biosurfaces: Plants, skin, and respiratory tract surfaces. Environ Health Perspect 1998;106(suppl 5): 1241–1251.
3 Pryor WA: Mechanisms of radical formation from reactions of ozone with target molecules in the lung. Free Radic Biol Med 1994;17:451–465.
4 Weber SU: Oxidants in skin pathophysiology; in Sen CK, Packer L, Hanninen O (eds): Exercise and Oxygen Toxicity, ed 2. Amsterdam, Elsevier Science, 1999.
5 Thiele JJ, Podda M, Packer L: Tropospheric ozone: An emerging environmental stress to skin. Biol Chem 1997;378:1299–1305.
6 Shindo Y, Witt E, Han D, Epstein W, Packer L: Enzymic and non-enzymic antioxidants in epidermis and dermis of human skin. J Invest Dermatol 1994;102:122–124.

7 Weber SU, Thiele JJ, Cross CE, Packer L: Communication: Vitamin C, uric acid, and glutathione gradients in murine stratum corneum and their susceptibility to ozone exposure (in process citation). J Invest Dermatol 1999;113:1128–1132.
8 Pryor WA: How far does ozone penetrate into the pulmonary air/tissue boundary before it reacts? Free Radic Biol Med 1992;12:83–88.
9 Thiele JJ, Traber MG, Podda M, Tsang K, Cross CE, Packer L: Ozone depletes tocopherols and tocotrienols topically applied to murine skin. FEBS Lett 1997;401:167–170.
10 Thiele JJ, Traber MG, Tsang K, Cross CE, Packer L: In vivo exposure to ozone depletes vitamins C and E and induces lipid peroxidation in epidermal layers of murine skin. Free Radi Biol Med 1997;23:385–391.
11 Weber SU, Jothi S, Thiele JJ: High pressure liquid chromatography analysis of ozone-induced depletion of hydrophilic and lipophilic antioxidants in murine skin. Methods Enzymol, in press.
12 Thiele JJ, Traber MG, Polefka TG, Cross CE, Packer L: Ozone-exposure depletes vitamin E and induces lipid peroxidation in murine stratum corneum. J Invest Dermatol 1997;108:753–757.
13 Moldau H, Laisk A: Ozone concentration in leaf intercellular air space is close to zero. Plant Physiol 1989;90:1163–1167.
14 Elias PM, Feingold KR: Lipids and the epidermal water barrier: Metabolism, regulation, and pathophysiology. Semin Dermatol 1992;11:176–182.
15 Uppu RM, Cueto R, Squadrito GL, Pryor WA: What does ozone react with at the air/lung interface? Model studies using human red blood cell membranes. Arch Biochem Biophys 1995;319:257–266.
16 Van der Vliet A, O'Neil CA, Eiserich JP, Cross CE: Oxidative damage to extracellular fluids by ozone and possible protective effects of thiols. Arch Biochem Biophys 1995;321:43–50.
17 Shindo Y, Witt E, Han D, Tzeng B, Aziz T, Nguyen L, Packer L: Recovery of antioxidants and reduction in lipid hydroperoxides in murine epidermis and dermis after acute ultraviolet radiation exposure. Photodermatol Photoimmunol Photomed 1994;10:183–191.
18 Buettner GR: The pecking order of free radicals and antioxidants: Lipid peroxidation, alpha-tocopherol, and ascorbate. Arch Biochem Biophys 1993;300:535–543.
19 Packer JE, Slater TF, Willson RL: Direct observation of a free radical interaction between vitamin E and vitamin C. Nature 1979;278:737–738.
20 Thiele JJ, Packer L: Noninvasive measurement of α-tocopherol gradients in human stratum corneum by high-performance liquid chromatography analysis of sequential tape strippings. Methods Enzymol 1999;300:413–419.
21 Esterbauer H, Striegl G, Puhl H, Rotheneder M: Continuous monitoring of in vitro oxidation of human low density lipoprotein. Free Radic Res Commun 1989;6:67–75.
22 Pryor WA, Das B, Church DF: The ozonation of unsaturated fatty acids: Aldehydes and hydrogen peroxide as products and possible mediators of ozone toxicity. Chem Res Toxicol 1991;4:341–348.
23 Roehm JN, Hadley JG, Menzel DB: Oxidation of unsaturated fatty acids by ozone and nitrogen dioxide: A common mechanism of action. Arch Environ Health 1971;23:142–148.
24 Pryor WA, Stanley JP, Blair E, Cullen GB: Autoxidation of polyunsaturated fatty acids. I. Effect of ozone on the autoxidation of neat methyl linoleate and methyl linolenate. Arch Environ Health 1976;31:201–210.
25 Thiele JJ, Traber MG, Re R, Espuno N, Yan LJ, Cross CE, Packer L: Macromolecular carbonyls in human stratum corneum: A biomarker for environmental oxidant exposure? FEBS Lett 1998;422:403–406.
26 Valacchi G, Weber SU, Luu C, Packer L: Ozone potentiates vitamin E depletion by ultraviolet radiation in the murine stratum corneum. FEBS Lett 2000;466:165–168.
27 Mao-Qiang M, Feingold KR, Thornfeldt CR, Elias PM: Optimization of physiological lipid mixtures for barrier repair. J Invest Dermatol 1996;106:1096–1101.
28 Ghadially R, Brown BE, Sequeira-Martin SM, Feingold KR, Elias PM: The aged epidermal permeability barrier: Structural, functional, and lipid biochemical abnormalities in humans and a senescent murine model. J Clin Invest 1995;95:2281–2290.
29 Proksch E, Feingold KR, Man MQ, Elias PM: Barrier function regulates epidermal DNA synthesis. J Clin Invest 1991;87:1668–1673.

30 Menon GK, Feingold KR, Elias PM: Lamellar body secretory response to barrier disruption. J Invest Dermatol 1992;98:279–289.
31 Wood LC, Jackson SM, Elias PM, Grunfeld C, Feingold KR: Cutaneous barrier perturbation stimulates cytokine production in the epidermis of mice. J Clin Invest 1992;90:482–487.
32 Wood LC, Feingold KR, Sequeira-Martin SM, Elias PM, Grunfeld C: Barrier function coordinately regulates epidermal IL-1 and IL-1 receptor antagonist mRNA levels. Exp Dermatol 1994;3:56–60.
33 Wood LC, Elias PM, Sequeira-Martin SM, Grunfeld C, Feingold KR: Occlusion lowers cytokine mRNA levels in essential fatty acid-deficient and normal mouse epidermis, but not after acute barrier disruption. J Invest Dermatol 1994;103:834–838.
34 Frampton MW, Pryor WA, Cueto R, Cox C, Morrow PE, Utell MJ: Ozone exposure increases aldehydes in epithelial lining fluid in human lung. Am J Respir Crit Care Med 1999;159:1134–1137.
35 Haddad EB, Salmon M, Koto H, Barnes PJ, Adcock I, Chung KF: Ozone induction of cytokine-induced neutrophil chemoattractant (CINC) and nuclear factor-kappa b in rat lung: Inhibition by corticosteroids. FEBS Lett 1996;379:265–268.

Stefan U. Weber, University of California, 251 Life Sciences Addition,
Berkeley, CA 94720–3200 (USA)
Tel. +1 (510) 642 7023, Fax +1 (510) 642 8313, E-Mail DrSWeber@yahoo.de

Thiele J, Elsner P (eds): Oxidants and Antioxidants in Cutaneous Biology.
Curr Probl Dermatol. Basel, Karger, 2001, vol 29, pp 62–73

Effects of UV and Visible Radiations on Cellular DNA

Jean Cadet, Thierry Douki, Jean-Pierre Pouget, Jean-Luc Ravanat, Sylvie Sauvaigo

Laboratoire 'Lésions des Acides Nucléiques', Département de Recherche Fondamentale sur la Matière Condensée/SCIB et UMR CNRS 5046, CEA/Grenoble, France

The effects of solar radiation on cellular DNA can be rationalized in terms of direct interactions of UVB radiation with pyrimidine and purine nucleobases on the one hand and photosensitization reactions mediated by UVA and visible light on the other [1–3]. However, it should be remembered that DNA bases can still absorb photons from the lower part of the UVA spectrum. As a result, cyclobutadipyrimidines (Pyr< >Pyr) that arise from direct excitation of either a cytosine or a thymine residue and a subsequent [2+2] cycloaddition of the latter reactive intermediate with an adjacent pyrimidine base are generated. Another important parameter to be considered is that UVA photons are able to penetrate more deeply into the epidermis layer than the higher-energetic UVB radiation, leading to a different distribution of DNA photodamage. Evidence is accumulating that the bulky dimeric pyrimidine photoproducts including Pyr< >Pyr, pyrimidine (6–4) pyrimidone photoadducts and related Dewar valence isomers are the main deleterious solar-radiation-induced lesions to DNA in terms of lethality, mutagenicity and carcinogenicity [4–8]. In this respect, it was found that adjacent cytosine residues constitute mutation hotspots in the p53 gene of human nonmelanoma skin cancers [9] and murine tumors [10], respectively, upon exposure to solar radiation. However, the dicytosine lesion(s) involved in the latter CC → TT transitions that may be considered as a molecular signature of exposure to sunlight remain to be identified. More recently, it has been postulated that UVA radiation may be carcinogenic through the induction of photooxidative DNA damage, even if there is little direct evidence to support such a proposal, at least, for human skin cells. The main objective of the present survey is to discuss the information that has recently become available on the formation of damage to cellular DNA upon exposure to the main components of solar radiation. This is mostly inferred from the application of sensitive methods that are able to measure the formation of direct excitation and oxidized photoproducts within cellular DNA. In this respect, the modified comet method that is rendered more specific by the use of DNA repair enzymes and the tandem mass spectrometry detection coupled to HPLC (HPLC-MS/MS) are among the most relevant and accurate available methods.

DNA Photoproducts Arising from Direct UVB Excitation

DNA photoproducts arise from excitation processes associated mostly with UVB radiation.

It has to be remembered that the discovery of the *cis-syn*-cyclobutadithymine (Thy< >Thy) in the early 60s has provided a strong impetus to the development of several fields of research including DNA photobiology but also, in a wider sense, genotoxicity and DNA repair. A large body of information is now available on the main direct photoreactions of DNA which in most cases are oxygen independent. In addition to the formation of cyclobutadipyrimidine (Pyr< >Pyr) already mentioned, excitation of a pyrimidine base in a sequence context where an adjacent pyrimidine residue including thymine, cytosine or 5-methylcytosine is present may lead to the induction of the corresponding pyrimidine (6–4) pyrimidone (Pyr-Pyo) photoadducts. Either a dioxetane or an azetidine instable intermediate is likely to be the precursor of the latter dimeric photoproducts when either a thymine or a cytosine residue are involved in the reaction on the 5′-end. An interesting photochemical feature of Pyr-Pyo photoadducts is their conversion into the related Dewar valence isomers upon UVB irradiation. However, it should be noted that the pyrimidine (6–4) pyrimidone adduct and the related Dewar valence isomer involving two adjacent cytosine bases have not yet been isolated and characterized in model studies, despite several attempts. Several other biadducts involving either adenine or thymine and adenine have been isolated and characterized [for reviews, see 1–3]. However, the quantum yields of their formation appear to be at least two orders of magnitude lower than those of Pyr< >Pyr and Pyr-Pyo. A large attention has been devoted in the past to the photo-induced formation of 6-hydroxy-5,6-dihydrocytosine, the so-called 'cytosine photohydrate'. This is the main monomeric DNA photodamage whose formation is accounted for by a singlet excited intermediate. More recently, it has been found that 8-oxo-7,8-dihydroguanine (8-oxo-Gua), a versatile but not specific biomarker of oxidative stress, is generated in isolated DNA upon exposure to UVB radiation. One possibility which has to be further confirmed is that 8-oxo-Gua arises from guanine singlet oxygen ($^1O_2$) oxidation. This is partly supported by the fact that the latter reactive oxygen species may be produced by energy transfer from triplet excited guanine and adenine residues with molecular oxygen. However, evidence was also provided for the occurrence of one-electron oxidation of the guanine moiety as the likely result of a charge transfer process with an excited base.

Various methods have been designed during the last three decades to monitor the formation of the main DNA dimeric pyrimidine photoproducts within isolated cells and tissues. The formation of *cis-syn* isomers of Thy< >Thy and also of the cyclobutadipyrimidines involving thymine and cytosine has been singled out by applying several chromatographic methods [for reviews, see 1, 3]. In this respect, HPLC on octadecylsilyl silica gel columns has shown its high capability. However, the weakness of the assays was the radioactive detection of the photolesions that required radiolabeling of cellular DNA. More indirect methods including the use of the UV radiation excision repair enzyme together with gel electrophoresis, alkaline elution or ligation-mediated polymerase chain reaction were also successfully developed [11]. Interestingly, the latter approach was used to map the formation of DNA photoproducts at the sequence level in genes of UV-irradiated cells. However, during the last 15 years, major achievements in the field of the measurement of DNA photodamage came from the availability of antibodies directed against the main classes of dimeric pyrimidine lesions. Thus, several immunological applications including location of Pyr< >Pyr within the epidermis of the skin [12] or the measurement of DNA dimeric lesions in isolated cells in association with either the flux cytometry technique or the comet assay [13] are available. In most cases, the detection of DNA photodamage was achieved by fluorescence measurement. Another interesting possibility provided by the antibodies is to monitor the formation of Pyr< >Pyr, Pyr-Pyo and other Dewar valence isomer photoproducts within extracted DNA using mostly ELISA methods. Overall, the immunological approach offers, with respect to the bulk of other methods, several major advantages such as versatility, sensitivity and easy application [for a review, see 1]. However, the immunoassays may suffer from a lack of specificity and also of appropriate calibration. The latter difficulties have recently been overcome with the availability of a powerful assay that associates highly resolutive HPLC with tandem mass spectrometry detection (HPLC-MS/MS). The ionization in the negative mode of the photoproducts [14] which are quantitatively released as dinucleoside monophosphates from DNA upon digestion with enzymes including 5′- and 3′-exonuclases together with nuclease P1 is achieved at atmospheric pressure using an electrospray source [15].

Interestingly, the formation of the twelve possible dimeric pyrimidine photoproducts including *cis-syn* Pyr< >Pyr, Pyr-Pyo and the related Dewar valence isomers can be simultaneously singled out in isolated DNA upon exposure to low doses of UVB radiation [15]. For this purpose, the sensitive multiple-reaction monitoring technique together with the negative mode detection was applied. This involves isolation of the molecular ion on the first quadrupole, followed by its fragmentation into the collision cell (second quadrupole) and subsequent analysis of a selected fragment on the third quadrupole. The process is highly specific, allowing the structural differentiation between a given Pyr< >Pyr and the related isomeric Pyr-Pyo photoadduct. Interestingly, the distribution pattern of the main UVC- and UVB-induced dimeric pyrimidine photoproducts is now available for isolated DNA. As a striking feature, we may notice a strong primary sequence dependence on the formation of both Pyr< >Pyr and Pyr-Pyo photoproducts. *cis-syn* Thy< >Thy is the main UVB photoproduct whereas cytosine dimeric damage including both Pyr< >Pyr and Pyo-Pyr measured as cyclobutadiuracil and 5-amino-5,6-dihydrouracil (6–4) pyrimidin-2-one, respectively, are rather generated in low yields. In addition, 5-hydroxy-5,6-dihydrouracil (6–4) 5-methylpyrimidin-2-one is the predominant Pyr-Pyo adduct at the cytosine-thymine sequence. The sensitivity of the assay which is close to a few femtomoles allows the accurate measurement of the bulk of dimeric pyrimidine lesions within cellular DNA upon low doses of UVC and UVB radiations [Douki and Cadet, unpubl. results]. A similar product distribution to that observed in isolated DNA was thus established with the predominance of the *cis-syn* Thy< >Thy. It was also confirmed that dicytosine residues are poorly photoreactive sites with respect to other dipyrimidine sequences as inferred from the low formation of the related cyclobutadipyrimidine and the lack of detection of the corresponding (6–4) photoadduct. Further development in the measurement of DNA photoproducts is expected from the association of the highly sensitive capillary gel electrophoresis method with the powerful electrospray ionization MS/MS detection technique. This is expected to provide an increase in the detection sensitivity by, at least, one order of magnitude with respect to that of the HPLC-MS/MS assay [15].

UVA- and Visible-Radiation-Mediated Oxidative Damage to DNA

Growing evidence indicates that oxidation reactions mediated by cellular photosensitizers are implicated in the deleterious effects of UVA and visible radiations on living systems. In this respect, unsaturated lipids and nucleic acids are the main targets of the photosensitization reactions which appear

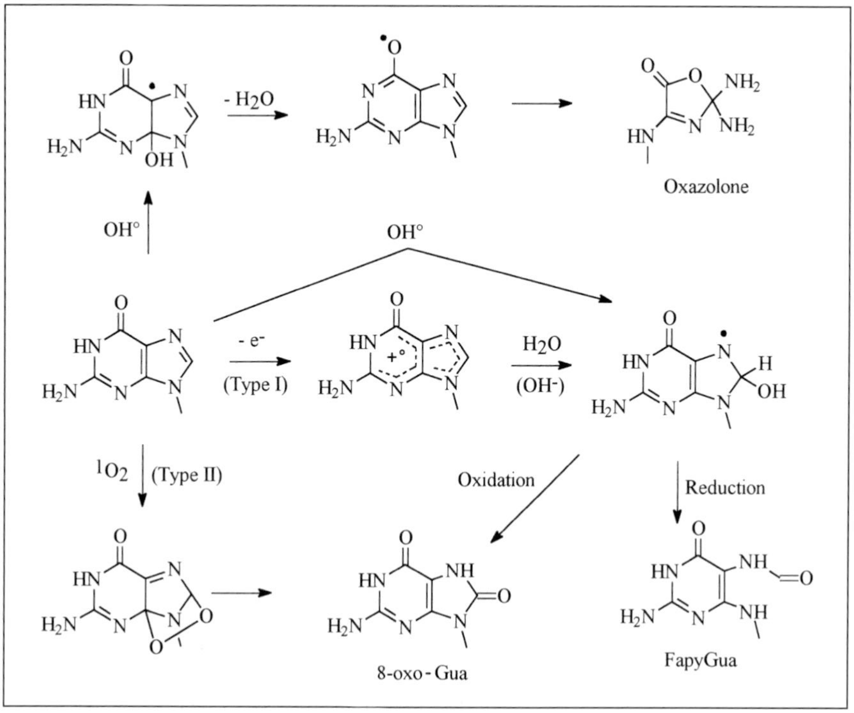

Fig. 1. Effects of solar radiation on cellular DNA.

to be in most cases oxygen dependent. A likely scenario implies that UVA- or visible-radiation-induced excitation of still unknown endogenous photosensitizers (quinones, flavins, porphyrins) is able to induce three main types of oxidative reactions. A photoexcited molecule may react with a substrate by two mechanisms. The so-called type I photosensitized process involves either one-electron transfer or hydrogen abstraction, most likely from the substrate to the photosensitizer (fig. 1). As a competitive process, whose relative contribution depends on the photosensitizer, energy transfer from the latter excited molecule to O_2 gives rise to singlet oxygen ($^1O_2$), a nonradical reactive oxygen species. In addition, superoxide radical ($O_2^{\bullet -}$) may be generated as a side product of a type I mechanism. The latter unreactive species is able to spontaneously or enzymatically dismutate into H_2O_2 which is the precursor of the highly reactive •OH radical generated in the Fe^{2+}-mediated Fenton reaction.

A large body of information is now available on the main oxidation reactions of isolated DNA and model compounds promoted by a one-electron process, $^1O_2$ and •OH radical, respectively [for recent reviews, see 16, 17]. The guanine base which exhibits the lowest ionization potential among the DNA constituents is the preferential target for one-electron oxidation reactions over adenine, pyrimidine bases and the sugar moiety (Gua > Ade > Thy – Cyt > 2-deoxyribose). The purine radical cation which arises from either the initial electron abstraction of a guanine residue or through hole transfer from a relatively distant one-electron oxidized purine or pyrimidine species is able to undergo two main reactions within double-stranded DNA. Hydration leads to the generation of the reducing 8-hydroxy-7,8-dihydroguanyl radical which is the precursor of 8-oxo-7,8-dihydroguanine (8-oxo-Gua) and 2,6-diamino-4-hydroxy-5-formamidopyrimidine (FapyGua). On the other hand, the competitive deprotonation reaction of the guanine radical cation was found to give rise, through a complex sequence of events, to 2,2,4-triamino-5(2H)-oxazolone (fig. 1) and its imidazolone precursor. Interestingly, •OH radical reactions with the guanine moiety produce similar degradation patterns. Addition of •OH at C-8 was shown to generate 8-oxo-Gua and FapyGua through the transient formation of the 8-hydroxy-7,8-dihydroguanyl radical. The other second major site of •OH addition is the C-4 position of the purine ring. Subsequent dehydration of the resulting radical intermediate is the critical step in the sequence of reactions giving rise to the oxazolone compound (fig. 1). In contrast to the rather unspecific •OH radical, which reacts almost equally with any organic molecules, $^1O_2$ is only able to oxidize guanine components. The main oxidation product of the reaction of the dienophile $^1O_2$ molecule within double-stranded DNA is 8-oxo-Gua (fig. 1). The latter oxidized DNA base appears to be a ubiquitous marker of oxidative stress since it is generated by, at least, four agents including peroxynitrite, •OH radical, $^1O_2$ and one-electron oxidants. The likely mechanism of the $^1O_2$ reaction involves a [4+2] Diels-Alder cycloaddition across the 4,8-bond of the purine ring and subsequent rearrangement of the resulting endoperoxides with the formation of 8-hydroperoxyguanine [18]. However, an almost completely different decomposition pattern of the initially generated 4,8-guanine endoperoxide was observed in nucleosides and short single-stranded oligonucleotides. Under the latter conditions, the 4*R* and 4*S* diastereoisomers of 4-hydroxy-8-oxo-4,8-dihydro-2′-deoxyguanosine (4-OH-8-oxo-dGuo), which can be used as specific $^1O_2$ markers, are predominantly produced [19]. Interestingly the measurement of the main one-electron (oxazolone and imidazolone) and singlet oxygen (4-OH-8-oxo-dGuo) can be used to assess the relative contribution of type I and type II processes to the overall photodynamic features of photosensitizers. Interestingly, it has recently been inferred from the results of a powerful HPLC/MS-MS measurement that riboflavin was a predominant type I photosensitizer whereas photoexcited methylene blue mostly acts through $^1O_2$ oxidation [20].

Measurement of UVA- and Visible-Light-Mediated Formation of Oxidized Bases and Strand Breaks within Cellular DNA

First it should be remembered that Pyr< >Pyr has been shown to be generated in a relatively low yield within cellular DNA upon exposure to UVA photons. It is likely that the latter UVA-mediated photodamage which does not contribute to the mutation spectrum of solar radiation (see below) arises from direct excitation of the pyrimidine bases and not from a triplet-triplet energy transfer process. However, one way to confirm this hypothesis would be the search for the lack of UVA induction of Pyr-Pyo photoadducts whose formation is usually accounted for by singlet excited intermediates. This could be achieved using the recently available HPLC/MS-MS assay (see above).

Major lines of evidence are accumulating for supporting the predominant role played by oxidation reactions in the effects of UVA and visible radiations within cells. Induction of heme oxygenase in fibroblast cells as a response to exposure to UVA radiation [21] is strongly suggestive of the occurrence of significant oxidative stress involving singlet oxygen and to a lesser extent of •OH radicals [22]. Interestingly, it has recently been shown that the induction of heme oxygenase participates in the protection against UVB immunosuppression [23]. It was also found that *L*-arginine, a substrate of nitric oxide synthase, was able to increase both genotoxic and cytotoxic effects of UVA radiation in keratinocytes [24]. Prooxidant effects of UVA radiation may be related, at least partly, to the immediate release of 'free' iron, a key component of the Fenton reaction, in irradiated fibroblast cells as the result of proteolysis of protein ferritin [25]. UVA has recently been found to promote photoaging of human fibroblasts in close association with mitochondrial common deletion through the participation of $^1O_2$ [26]. Another major piece of relevant information deals with the observed defense systems of skin against the UVA-induced reactive oxygen species provided by various antioxidant enzymes and reductants [27]. Interestingly, it was found that UVA radiation induces different mutation spectra than UVB photons in the episomal *LacZ'* gene of 293 human epithelial cells [28, 29]. One major feature was that UVA did not induce mutations at dipyrimidine sites. On the other hand, the observation of a higher incidence of G→T transversions is highly suggestive of the occurrence of oxidative reactions. This is likely due to the UVA-mediated generation of 8-oxo-Gua in the *aprt* locus of Chinese hamster ovary cells [5] and human skin tumors [30]. Indirect support for the significant involvement of 8-oxo-Gua in the deleterious effects of UVA radiation on microorganisms was inferred from the observed increase in G to T transversions in *Escherichia coli* strains deficient in Fpg protein [31, 32]. It should be, however, mentioned that the contribution of oxidative processes is, at best, a minor process in the induction of p53 within skin tumors of hairless mice upon UVA exposure [33]. More direct evidence for the induction of oxidative damage to cellular DNA by either UVA radiation or visible light was gained in recent years. Thus, the level of 8-oxo-Gua was found to increase within cellular DNA [34–39] and RNA [37] upon exposure to UVA radiation as inferred from HPLC electrochemical detection (ECD) measurements. In addition, relevant information was gained on the pattern distribution of major classes of DNA photodamage including dimeric pyrimidine photoproducts and oxidative lesions in UVA-irradiated L1210 mouse leukemia and AS52 Chinese hamster cells [40, 41]. This was inferred from the application of the highly sensitive alkaline elution assay associated with specific DNA repair enzymes. Visible light was found in both cases to induce the formation of Fpg-sensitive sites in relatively high yields by comparison with other DNA lesions including single-strand breaks, abasic residues, endonuclease-III-sensitive sites and Pyr< >Pyr. The cellular DNA damage profiles that were generated by visible light were similar to those produced under cell-free conditions by various photosensitizers acting through type I and type II mechanisms [42]. In addition, relevant information was gained from the consideration of the action spectra obtained for AS52 cells over the UV-visible range. Interestingly, we may note a parallel decrease in the level of single-strand breaks, Pyr< >Pyr and Fpg-sensitive sites within the 290- to 315-nm range. In addition, the level of T_4 endonuclease V sites (mostly Pyr< >Pyr) was found to continue to decrease in an exponential manner at longer wavelengths [43]. As striking features, the number of oxidized base residues, mostly 8-oxo-Gua, recognized by the Fpg protein showed a minimum around 340 nm and a maximum between 400 and 500 nm. The overall observations are likely to be accounted for by the overwhelming formation of 8-oxo-Gua as the result of predominant singlet oxygen oxidation. It may be added that the involvement of the •OH radical in the induction of Fpg-sensitive sites is, at best, a minor process, as inferred from the relatively low yields of oxidized pyrimidine bases and DNA strand breaks, the expected degradation products of this highly reactive oxygen species. This has received further confirmation by a recent detailed study of UVA-mediated formation of photooxidative lesions in the DNA of human THP-1 monocytes [44]. For this purpose, several biochemical and chemical assays were applied. The level of DNA strand breaks together with the number of Fpg- and endonuclease-III-sensitive sites was assessed using the conventional and modified comet assays [45, 46]. In addition, search for the photo-induced formation of 8-oxo-dGuo and FapyGua was performed by applying HPLC/ECD and HPLC/gas chromatography mass spectrometry. It should be added that the modified comet assay associated with Fpg, which was utilized for measuring 8-oxo-dGuo, was calibrated with determinations carried out with the HPLC/ECD method. Furthermore, the results of the UVA irradiation were compared with those obtained by exposing the cells to gamma rays, conditions under which the damage arises, at least, partly from radiation-induced •OH radicals. Interestingly, it was found that FapyGua, which is produced in about 2.5-fold higher yield than 8-oxo-Gua, upon exposure of cellular DNA to gamma rays is not generated, at least, in detectable amounts upon UVA irradiation. Another interesting information deals with the comparison of the ratio of 8-oxo-Gua/endonuclease III sites and 8-oxo-Gua/strand breaks which were between 16 and 20 times higher for UVA exposure compared to gamma irradiation. This may be rationalized in terms of the predominant role of $^1O_2$ in the oxidation reactions to the guanine moiety of cellular DNA. It should be added that the contribution of •OH radicals, which mostly explains the formation of DNA strand breaks and endonuclease-III-sensitive sites, is rather low. Another major remark deals with the fact that the formation of 8-oxo-Gua, the main UVA-induced lesion so far characterized in cellular DNA, namely 0.00098 8-oxo-Gua per 10^6 bases and per kilojoule per square meter of UVA radiation, is rather low. This means that the UVA dose exposure required to produce the equivalent of the cellular level background of 8-oxo-Gua is close to 80 kJ/m^2.

Conclusions and Future Directions

Over the last decade, significant progress has been made for a better understanding of the molecular effects of solar radiation on cellular DNA in terms of mechanistic features and biological endpoints. A major achievement deals with the recent development of the powerful HPLC-MS/MS method aimed at individually measuring relatively low levels of each of the twelve main dimeric pyrimidine photoproducts. This should allow detailed kinetic studies of the repair of the different lesions together with a better assessment of their biological role, with emphasis on photoproducts at dicytosine sites. The expected improvement in the detection sensitivity of the method by substituting the capillary gel electrophoretic analysis to the HPLC separation should promote the development of in vivo assays. The measurement of DNA photodamage in skin could be used as marker of exposure and for photoprotection studies. Interestingly, the currently available HPLC-MS/MS method could also be used to monitor the release of dimeric pyrimidine photoproducts in urine as the result of DNA repair through the nucleotide excision pathway.

One of the major messages left by recent studies concerning the molecular effects of UVA radiation on cellular DNA deals with the relatively low efficiency for near-UV photons to generate oxidative base damage through photosensitization reactions. Efforts have to be made to clearly identify the endogenous photosensitizers involved in the UVA- and visible-light-mediated oxidative stress. A related matter deals with a better assessment of the respective contribution of the two main reactive oxygen species, namely $^1O_2$ and to a lesser extent •OH radical, which are implicated in the overall oxidative damage to cellular DNA. The ratio is likely to vary with the investigated cell systems in relation with different qualitative and quantitative distributions of endogenous photosensitizers. Another lacking information concerns the structure of photosensitized DNA-protein cross-links which may represent major UVA-induced oxidative damage to DNA. There is also a growing interest for delineating the putative involvement of a still not yet well investigated class of DNA lesions that consists of adducts between amino-substituted nucleobases and breakdown products of lipid peroxidation. In this respect, propano and etheno adducts that arise from the addition of malonaldehyde and 4-hydroxynonenal with purine bases are likely to be interesting candidates.

References

1 Cadet J, Vigny P: The photochemistry of nucleic acids; in Morrison H (ed): Bioorganic Photochemistry. New York, Wiley & Sons, 1990, vol 1, pp 1–272.
2 Taylor JS: DNA, sunlight and skin cancer. Pure Appl Chem 1995;67:183–190.
3 Cadet J, Berger M, Douki T, Morin B, Raoul S, Ravanat JL, Spinelli S: Effects of UV and visible radiation on DNA – Final base damage. Biol Chem 1997;378:1275–1286.
4 Clingen PH, Arlett CF, Cole J, Wauch APW, Lowe JE, Harcourt SA, Hermanova N, Roza L, Mori T, Nikaido O, Green MHL: Correlation of UVC and UVB cytotoxicity with the induction of specific photoproducts in T-lymphocytes and fibroblasts from normal human donors. Photochem Photobiol 1995;61:163–170.
5 Drobetsky EA, Turcotte J, Chateauneuf A: A role for ultraviolet A in solar mutagenesis. Proc Natl Acad Sci USA 1995;92:2350–2354.
6 Cleaver J: Mutagenic lesions in photocarcinogenesis: The fate of pyrimidine photoproducts in repair-deficient disorders. Photochem Photobiol 1996;63:377–379.
7 Sage E, Lamolet B, Brulay E, Moustacchi E, Chateauneuf A, Drobetsky EA: Mutagenic specificity of solar UV light in nucleotide excision repair-deficient rodent cells. Proc Natl Acad Sci USA 1996; 93:176–180.
8 Sarasin A: The molecular pathways of ultraviolet carcinogenesis. Mutat Res 1999;428:5–10.
9 Ziegler A, Leffell DJ, Kunala S, Sharma HW, Gailani M, Simon JA, Halperin AJ, Baden HP, Shapiro PE, Bale AE, Brash DE: Mutation hotspots due to sunlight in the p53 gene of non-melanoma skin cancers. Proc Natl Acad Sci USA 1993;90:4216–4220.
10 Kanjilal S, Piercall WE, Cummings KK, Kripke ML, Ananthaswamy HN: High frequency of p53 mutations in ultraviolet radiation-induced murine skin tumors: Evidence for strand bias and tumor heterogeneity. Cancer Res 1993;53:2961–2963.
11 Pfeifer GP, Drouin R, Holmquist GP: Detection of DNA adducts at the DNA sequence level by ligation-mediated PCR. Mutat Res 1993;288:39–46.
12 Hattori Y, Nishigori C, Tanaka T, Uchida K, Nikaido O, Osawa T, Hiai H, Imamura S, Toyokuni S: 8-Hydroxy-2′-deoxyguanosine is increased in epidermal cells of hairless mice after ultraviolet B exposure. J Invest Dermatol 1997;107:733–737.

13 Sauvaigo S, Serres C, Signorini N, Emonet N, Richard MJ, Cadet J: Use of single-cell gel electrophoresis assay for the immunofluorescent detection of specific DNA damage. Anal Biochem 1998;259: 1–7.
14 Douki T, Court M, Cadet J: Electrospray-mass spectrometry characterization and measurement of far-UV induced thymine photoproducts. J Photochem Photobiol B 2000;54:145–154.
15 Douki T, Court M, Sauvaigo S, Odin F, Cadet J: Formation of the main UV-induced thymine dimeric lesions within isolated and cellular DNA as measured by high performance liquid chromatography-tandem mass spectrometry. J Biol Chem 2000;275:11678–11685.
16 Cadet J, Berger M, Douki T, Ravanat JL: Oxidative damage to DNA: Formation, measurement, and biological significance. Rev Physiol Biochem Pharmacol 1997;131:1–87.
17 Cadet J, Delatour T, Douki T, Gasparutto D, Pouget JP, Ravanat JL, Sauvaigo S: Hydroxyl radicals and DNA base damage. Mutat Res 1999;424:9–21.
18 Sheu C, Foote CS: Endoperoxide formation in a guanosine derivative. J Am Chem Soc 1993;115: 10446–10447.
19 Ravanat JL, Cadet J: Reaction of singlet oxygen with 2′-deoxyguanosine and DNA: Isolation and characterization of the main oxidation products. Chem Res Toxicol 1995;8:379–388.
20 Ravanat JL, Remaud G, Cadet J: Measurement of the main photooxidation products of 2′-deoxyguanosine using chromatographic methods coupled to mass spectrometry. Arch Biochem Biophys 2000;374:118–127.
21 Vile GF, Tyrrell RM: Oxidative stress resulting from ultraviolet A irradiation of human skin fibroblasts leads to a heme oxygenase-dependent increase in ferritin. J Biol Chem 1993;268:14678–14681.
22 Basu-Modak S, Tyrrell RM: Singlet oxygen: A primary effector in the ultraviolet A/near-visible light induction of the human oxygenase gene. Cancer Res 1993;53:4505–4510.
23 Reeve VE, Tyrrell RM: Heme oxygenase induction mediates the photoimmunoprotective activity of UVA radiation in the mouse. Proc Natl Acad Sci USA 1999;96:9317–9323.
24 Didier C, Emonet-Piccardi N, Béani JC, Cadet J, Richard MJ: *L*-Arginine increases UVA cytotoxicity in irradiated human keratinocyte cell line: Potential role of nitric oxide. FASEB J 1999;13:1817–1824.
25 Pourzard C, Watkin RD, Brown JE, Tyrrell RM: Ultraviolet A radiation induces immediate release of iron in human primary skin fibroblasts: The role of ferritin. Proc Natl Acad Sci USA 1999;96: 6751–6756.
26 Berneburg M, Grether-Beck S, Kürten V, Ruzicka T, Briviba K, Sies H, Krutmann J: Singlet oxygen mediates the UVA-induced generation of the photoaging-associated mitochondrial common deletion. J Biol Chem 1999;274:15345–15349.
27 Applegate LA, Frenk E: Cellular defence mechanism of the skin against oxidant stress and in particular UVA radiation. Eur J Dermatol 1995;5:97–103.
28 Robert C, Muel B, Benoit A, Dubertret L, Sarasin A, Stary A: Cell survival and shuttle vector mutagenesis induced by ultraviolet A and ultraviolet B radiation in a human cell line. J Invest Dermatol 1996;106:721–728.
29 Stary A, Robert C, Sarasin A: Deleterious effects of ultraviolet A radiation in human cells. Mutat Res 1997;383:1–8.
30 Daya-Grosjean L, Dumaz N, Sarasin A: The specificity of p53 mutation spectra in sunlight induced human cancers. J Photochem Photobiol B Biol 1995;28:115–124.
31 Shennan MG, Palmer CM, Schellhorn HE: Role of Fapy glycosylase and UvrABC excinuclease in the repair of UVA (320–400 nm) mediated DNA damage. Photochem Photobiol 1996;63:68–73.
32 Palmer CM, Serafini DM, Schellhorn HE: Near ultraviolet radiation (UVA and UVB) causes a formamidopyrimidine glycosylase dependent increase in G to T transversions. Photochem Photobiol 1997;65:543–549.
33 van Kranen HJ, de Laat A, van de Ven J, Wester PW, de Vries A, Berg RJW, van Kreijl CF, de Gruijl FR: Low incidence of p53 mutations in UVA (365 nm)-induced skin tumors in hairless mice. Cancer Res 1997;57:1238–1240.
34 Fischer-Nielsen A, Loft S, Jensen KG: Effect of ascorbate and 5-aminosalicylic acid on light-induced 8-hydroxydeoxyguanosine formation in V79 Chinese hamster cells. Carcinogenesis 1993; 14:2431–2433.

35 Rosen JE, Prahalad AK, Williams GM: 8–Oxodeoxyguanosine formation in the DNA of cultured cells after exposure to H_2O_2 alone or with UVB or UVA irradiation. Photochem Photobiol 1996; 64:117–122.
36 Kvam E, Tyrrell RM: Induction of oxidative damage in human skin cells by UV and near visible radiation. Carcinogenesis 1997;17:2379–2384.
37 Wamer WG, Wei RR: In vitro photooxidation of nucleic acids by ultraviolet A radiation. Photochem Photobiol 1997;65:560–563.
38 Zhang X, Rosenstein BS, Wang Y, Lebwohl M, Mitchell DM, Wei H: Induction of 8-oxo-7,8-dihydro-2′-deoxyguanosine by ultraviolet radiation in calf thymus DNA and HeLa cells. Photochem Photobiol 1997;65:119–124.
39 Douki T, Perdiz D, Frof P, Kuluncsics Z, Moustacchi E, Cadet J, Sage E: Oxidation of guanine in cellular DNA by solar UV radiation: Biological role. Photochem Photobiol 1999;70:184–190.
40 Pflaum M, Boiteux S, Epe B: Visible light generates oxidative DNA base modifications in large excess of strand breaks in mammalian cells. Carcinogenesis 1994;15:297–300.
41 Pflaum M, Kielbassa C, Gramyn M, Epe B: Oxidative DNA damage induced by visible light in mammalian cells: Extent, inhibition by antioxidants and genotoxic effects. Mutat Res 1998;408: 137–146.
42 Epe B: DNA damage profiles induced by oxidizing agents. Rev Physiol Biochem Pharmacol 1995; 127:223–233.
43 Kielbassa C, Roza L, Epe B: Wavelength dependence of oxidative DNA damage induced by UV and visible light. Carcinogenesis 1997;18:811–816.
44 Pouget JP, Douki T, Richard MJ, Cadet J: DNA damage induced in cells by gamma and UVA radiations as measured by HPLC/GC-MS, HPLC-EC and comet assays. Chem Res Toxicol 2000; 13:541–549.
45 Collins AR, Dusinska M, Gedik CM, Stetina R: Oxidative damage to DNA: Do we have a reliable biomarker? Environ Health Perspect 1996;104:465–469.
46 Pouget JP, Ravanat JL, Douki T, Richard MJ, Cadet J: Measurement of DNA base damage in cells exposed to low doses of gamma radiation: Comparison between the HPLC-EC and the comet assays. Int J Radiat Biol 1999;75:51–58.

Dr. Jean Cadet, Laboratoire 'Lésions des Acides Nucléiques', DRFMC/SCIB, CEA/Grenoble, F–38054 Grenoble Cedex 9 (France)
Tel. +33 4 76 88 49 87, Fax +33 4 76 88 50 90, E-Mail cadet@drfmc.ceng.cea.fr

Thiele J, Elsner P (eds): Oxidants and Antioxidants in Cutaneous Biology.
Curr Probl Dermatol. Basel, Karger, 2001, vol 29, pp 74–82

Sequence-Specific DNA Damage Induced by UVA Radiation in the Presence of Endogenous and Exogenous Photosensitizers

Shosuke Kawanishi, Yusuke Hiraku

Department of Hygiene, Mie University School of Medicine, Tsu, Mie, Japan

Solar radiation is the main source of exposure to UV radiation and sufficient evidence for its carcinogenicity in humans has been provided [1]. Solar UV radiation reaching the Earth consists of UVA (320–380 nm) and UVB (290–320 nm). UVB has been believed to play the key role in solar carcinogenesis. UVB is directly absorbed by the DNA molecule to form cyclobutane pyrimidine dimers and pyrimidine (6–4) pyrimidone photoadducts, resulting in mutation and carcinogenesis. Recent studies on UV carcinogenesis have revealed that UVA is also mutagenic and carcinogenic [2]. Although the carcinogenic potential of UVA is much smaller than that of UVB, UVA, which comprises approximately 95% of solar UV radiation [1], may play an important role in solar carcinogenesis. A study on the mutagenic specificity of solar radiation has indicated that not only UVB, but also UVA, participates in solar mutagenesis [2]. In contrast to UVB, UVA indirectly induces DNA damage by activation of photosensitizers, because UVA is hardly absorbed by the DNA molecule and forms little or no DNA photoadducts. Therefore, solar carcinogenesis involves UVA-mediated photoactivation of endogenous photosensitizers. We have demonstrated that a variety of endogenous photosensitizers (e.g. porphyrins and flavins) mediate UVA-induced DNA damage [3–5].

In addition to endogenous molecules, certain drugs may act as exogenous photocarcinogens. Psolarens have been used for the treatment of skin diseases, particularly psoriasis, in combination with UVA irradiation (PUVA therapy), and it is known that the incidence of skin tumors is increased by PUVA therapy

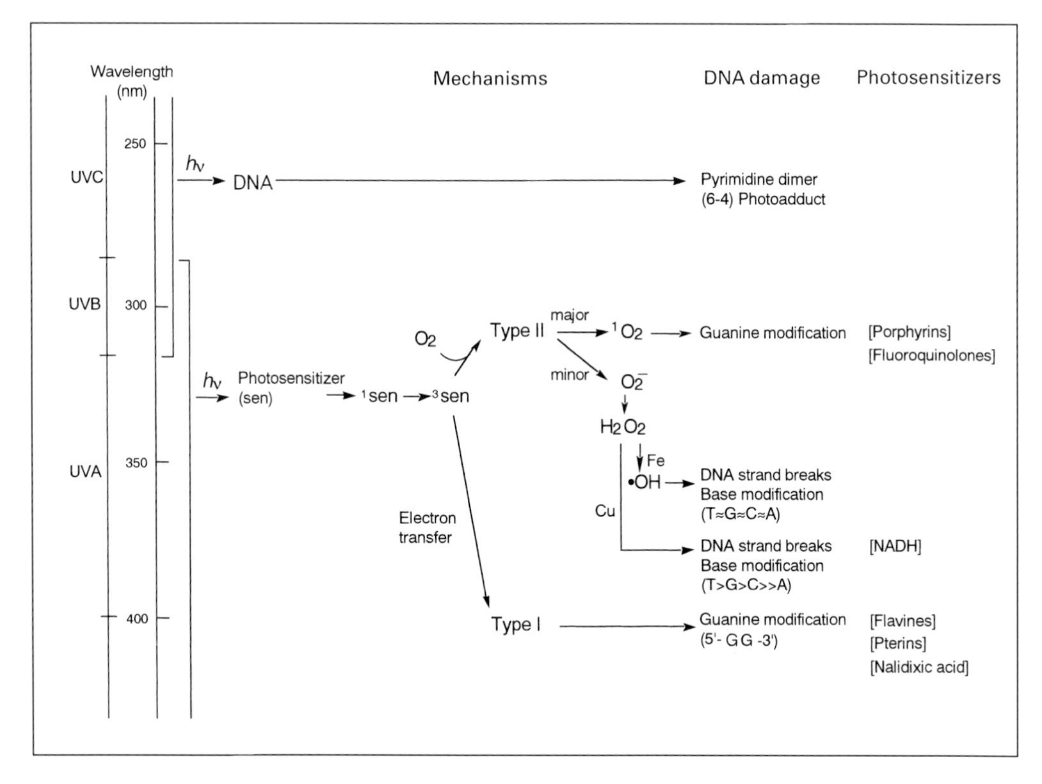

Fig. 1. Mechanisms of UV-induced DNA damage. Photosensitizers tested in our studies are listed in brackets.

[1]. Recently, it has been reported that nalidixic acid (NA) and fluoroquinolone antibacterials, which have been widely used for the treatment of infectious diseases, cause skin tumors in animals exposed to UVA, and lomefloxacin (LFLX) showed the most potent photocarcinogenic effect among the drugs tested [6, 7]. Therefore, these studies suggest that these drugs may have photocarcinogenic effects. We have demonstrated that NA [8] and fluoroquinolones [unpubl. data] mediate UVA-induced DNA damage.

UVA radiation causes DNA damage through type I (electron transfer), major type II [singlet oxygen ($^1O_2$)] and minor type II [superoxide anion radical (O_2^-)] mechanisms. The type I mechanism does not require oxygen for the induction of DNA damage, whereas type II mechanisms proceed only in the presence of oxygen (fig. 1). The mechanisms are dependent on the chemical properties of photosensitizers. The mechanisms and site specificity of UV-induced DNA photodamage are discussed below.

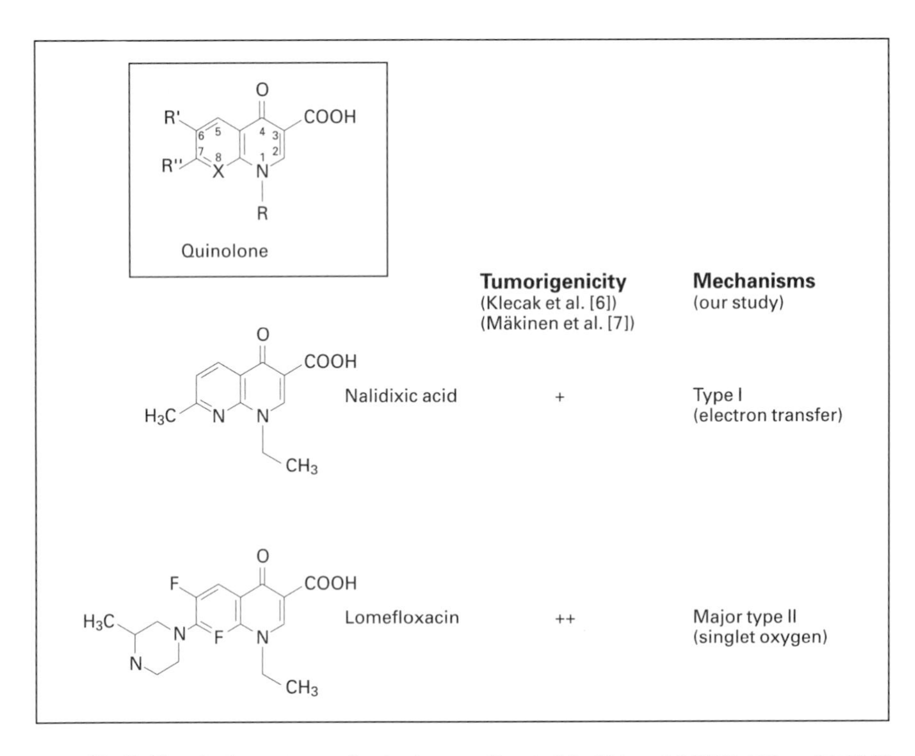

Fig. 2. Chemical structures of quinolone antibacterials, NA and LFLX. NA and LFLX show tumorigenic effects upon UVA irradiation and cause DNA damage via type I and major type II mechanisms, respectively.

Type I Mechanism

The type I mechanism involves electron transfer through the interaction of an excited photosensitizer with a DNA base. This mechanism is dependent on the oxidation potential of the DNA base and the reduction potential of the excited photosensitizer. Guanine has the lowest oxidation potential among the four DNA bases, that is guanine is most likely to be oxidized. We have demonstrated that various endogenous molecules, such as riboflavin [3] and pterin derivatives [4], act as photosensitizers mediating DNA damage through electron transfer when double-stranded DNA is exposed to 365-nm UVA light in the presence of these molecules.

Recently, we have focused on a quinolone antibacterial, NA, as an exogenous photosensitizer. The chemical structure of NA is shown in figure 2. NA has been used for the treatment of urinary tract infections. Animal experiments demonstrated that NA caused skin tumors in mice exposed to UVA [6, 7] and suggested that NA could act as an exogenous photocarcinogen. We have found that NA causes damage to DNA fragments obtained from the human c-Ha-*ras*-1 protooncogene and the p353 tumor suppressor gene upon UVA irradiation [8]. DNA damage was observed only when DNA fragments were treated with piperidine, suggesting that the damage is due to base modification with little or no strand breaks. NA caused photodamage in double-stranded DNA at consecutive guanines, particularly at the 5′-G in the 5′-GG-3′ sequence, while little or no damage was observed at single guanines. In single-stranded DNA, guanines were specifically damaged, but the site specificity for consecutive guanines was not observed. The measurement of 8-oxo-7,8-dihydro-2′-deoxyguanosine (8-oxo-dG) with an HPLC coupled with an electrochemical detector showed that 8-oxo-dG was formed more efficiently in double-stranded DNA than in single-stranded DNA. These results suggest that the double-helical structure contributes to efficient NA-induced DNA photodamage and determination of its site specificity. The electron spin resonance spin destruction method showed that the addition of dGMP to photoexcited NA rapidly quenched the signal of the 2,2,6,6-tetramethyl-4-piperidone-N-oxyl (4-oxo-TEMPO) radical, whereas dTMP, dAMP and dCMP had little or no effect on the signal. This finding suggests that 1 electron is transferred from guanine to photoexcited NA, resulting in the formation of the NA anion radical and the guanine cation radical, which interact with the 4-oxo-TEMPO radical to attenuate the signal [8].

On the basis of these findings, we proposed the mechanisms of guanine oxidation in GG and GGG sequences through electron transfer as shown in figure 3. The energy level of the highest occupied molecular orbital (HOMO) of guanine is highest among the four DNA bases, and therefore guanine is most likely to be oxidized. Recently, more detailed calculations have revealed that a large part of HOMO is concentrated and electron loss centers are localized on the 5′-G of GG doublets in the B-form double-stranded DNA, and that stacking of two guanine bases significantly lowers the ionization potential [9]. Therefore, electron transfer occurs specifically at this site to produce the guanine cation radical. In GGG triplets, HOMO is mainly distributed on the 5′-G, and therefore this guanine is easily oxidized by electron transfer. However, the guanine cation radical formed at the 5′-G is reduced by electron transfer from the middle guanine (hole migration), because the radical on the middle G is estimated to be the most stable in certain GGG triplets [10]. Therefore, the middle G is most likely to be damaged. The guanine cation radical produced by electron transfer undergoes hydration, followed by subsequent oxidation to 8-oxo-dG [11]. Recently, a more detailed investigation has revealed that the selectivity of GGG triplets towards DNA photodamage depends on the flanking bases. Riboflavin and a benzophenone derivative induced DNA photodamage specifically at GGG triplets, but different selecti-

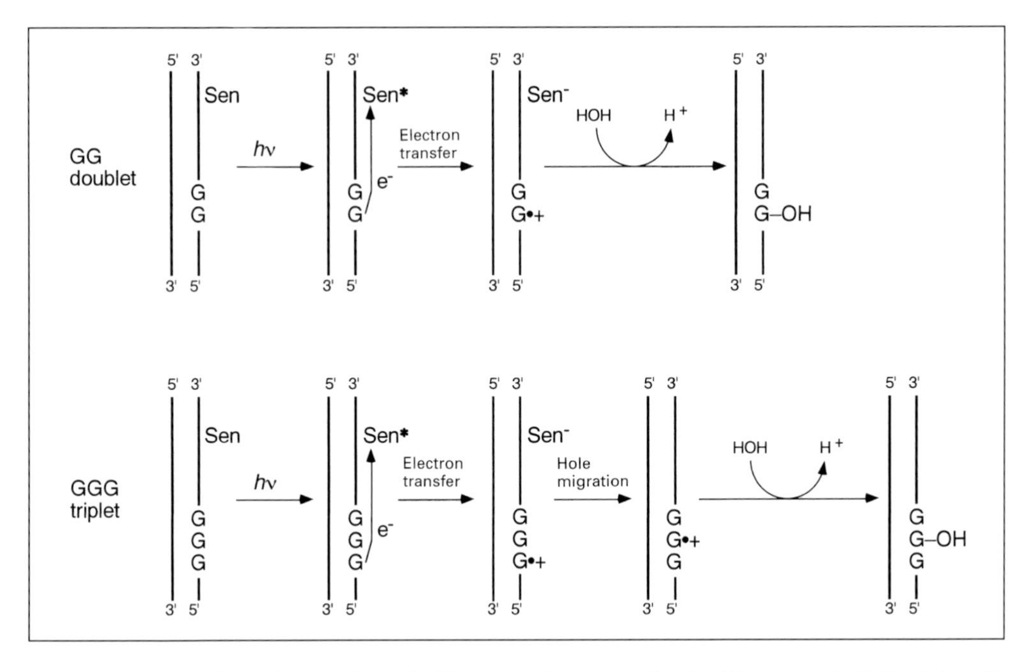

Fig. 3. Mechanisms of DNA photodamage at GG doublets (***a***) and GGG triplets (***b***) by electron transfer. Sen = Photosensitizer; Sen* = excited photosensitizer.

vity was observed in different sequences: the 5′-G and the middle G were specifically damaged in 5′-CGGGC-3′ and 5′-TGGGT-3′ sequences, respectively [10]. Oxidative damage to guanines mediated by electron transfer can be followed by long-range electron transfer leading to oxidation of guanines at remote sites where a large part of HOMO is distributed. Rhodium and ruthenium compounds tethered to the DNA duplex induced DNA photodamage at the 5′-G of the 5′-GG-3′ sequence by long-range electron transfer [12, 13]. Recently, it has been reported that guanine oxidation is induced by DNA-mediated charge transport over 200 Å [13]. These studies demonstrated that the DNA double helix could serve as a medium for charge migration over long distances in UVA-mediated guanine oxidation.

Our findings on the site specificity of DNA damage and 8-oxo-dG formation suggest that 8-oxo-dG is formed specifically at the 5′-G of GG doublets through electron transfer. In addition to 8-oxo-dG, imidazolone and oxazolone are generated from the guanine cation radical through deprotonation and oxidation [11]. The formation of these oxidative products of guanine by a type I mechanism would play the key role in carcinogenesis induced by endogenous and exogenous photosensitizers.

Major Type II Mechanism

The major type II mechanism involves energy transfer from an excited photosensitizer to molecular oxygen to produce $^1O_2$, which is a very powerful oxidant with a relatively long lifetime and which reacts with many cellular macromolecules including DNA. A variety of endogenous and exogenous compounds have been shown to generate $^1O_2$ in response to UVA irradiation and cause DNA damage specifically at guanines. The interaction of guanine with $^1O_2$ leads to the formation of 8-hydroperoxyguanine, followed by subsequent reduction to 8-oxo-dG [11].

Hematoporphyrin has been reported to induce DNA damage specifically at guanine residues through the generation of $^1O_2$ [3, 14]. DNA damage and 8-oxo-dG formation induced by hematoporphyrin + UVA were enhanced in D_2O, suggesting that hematoporphyrin-induced DNA photolesions involve the generation of $^1O_2$ [3]. These findings suggest that porphyrin derivatives may act as endogenous photosensitizers and participate in solar carcinogenesis.

Exogenous molecules may cause cancer through DNA photodamage in a similar mechanism. Fluoroquinolone antibacterials, which contain a fluorine at the C-6 position of the quinolone nucleus (fig. 2), have recently been used for a variety of infectious diseases because of their much broader antimicrobial activities than those of quinolones including NA. Fluoroquinolones have been reported to cause skin tumors in animals exposed to UVA [6, 7]. Fluoroquinolines (LFLX, fleroxacin, ciprofloxacin and ofloxacin) showed more potent tumorigenic effects than NA. LFLX formed a number of squamous-cell carcinomas in animals whereas other fluoroquinolones formed no or few malignant tumors [6, 7]. LFLX is distributed to plasma and skin more efficiently than other fluoroquinolones [6]. Therefore, these fluoroquinolones, particularly LFLX, could act as exogenous photocarcinogens. Our preliminary experiment showed that LFLX caused DNA damage specifically at guanines by generating $^1O_2$ [unpubl. data]. Fluoroquinolones may exhibit their photocarcinogenicity through DNA damage induced by a similar mechanism. A study on structure-side-effect relationship revealed that the degree of phototoxicity exhibited by fluoroquinolones is partly attributed to the molecular moiety at the 8-position of the fluoroquinolone nucleus [15]. The 8-position substituent with halogen (fluorine or chlorine) causes the greatest photoreaction among the drugs tested [15].

The reaction of guanine with $^1O_2$ leads to the formation of 8-hydroperoxyguanine and the subsequent reduction to 8-oxo-dG [10]. Furthermore, $^1O_2$ oxidizes 8-oxo-dG to various oxidative products including imidazolone and oxazolone [11]. Certain endogenous and exogenous photosensitizers mediate the formation of these oxidative products by a type II major mechanism, leading to photomutagenesis and photocarcinogenesis.

Minor Type II Mechanism

The minor type II mechanism is mediated by the formation of O_2^- by electron transfer from an excited photosensitizer to molecular oxygen, followed by dismutation to hydrogen peroxide (H_2O_2). O_2^- may also be generated by the interaction of oxygen with the sensitizer anion radical produced by the type I mechanism. Although O_2^- and H_2O_2 are not capable of causing DNA damage by themselves, H_2O_2 can cause DNA damage in the presence of metal ions. Free hydroxyl radical (•OH) is generated by the reaction of H_2O_2 with Fe(II) ion (the Fenton reaction). •OH is known to cause DNA damage at every nucleotide with little or no site selectivity [16]. In contrast, H_2O_2 induces site-specific DNA damage particularly at thymine and guanine residues in the presence of Cu(II), and the primary reactive species causing DNA damage appears to be copper-oxygen complexes [17].

It has been demonstrated that sepiapterin, an endogenous dihydropterin, induced DNA photodamage at the 5′-G of the 5′-GG-3′ sequences, whereas in the presence of Cu(II), DNA damage was induced preferably at thymine of the 5′-GTC-3′ sequence [4]. Catalase and bathocuproine, a Cu(I)-specific chelating agent, inhibited DNA photolesions, suggesting that DNA damage involved H_2O_2 and Cu(I). It is speculated that photoexcited sepiapterin reacts with oxygen to form O_2^-, which is dismutated to H_2O_2, and reactive species generated from the reaction of H_2O_2 with Cu(I) played the key role in DNA photodamage [4]. A similar result was observed with NADH, an endogenous reductant [unpubl. data].

Role of UVA-Induced Oxidative DNA Damage in Mutagenesis and Carcinogenesis

The mechanisms of DNA damage induced by UVA radiation in the presence of a variety of endogenous and exogenous photosensitizers have been extensively studied. The mechanisms of DNA damage are dependent on photosensitizers and determine the site specificity (fig. 1). In the type I mechanism, DNA damage is induced specifically at the 5′-G of 5′-GG-3′ sequences, whereas the major type II mechanism mediates damage to most guanine residues without specificity for consecutive guanines. It is presumed that a large part of damaged guanines is accounted for by 8-oxo-dG. The formation of 8-oxo-dG through type I and major type II mechanisms may lead to DNA misreplication, resulting in mutation, particularly G→T transversions [18]. The mutations at consecutive guanines, such as GGT→TGT and GGC→TGC, in *ras* oncogenes were observed in human skin cancer [19]. These mutations may be induced by 8-oxo-dG formation at the 5′-G of the 5′-GG-3′ sequence by a type I mechanism. DNA damage induced by a minor type II mechanism is dependent on metal ions. Our studies have shown that reactive species generated in the presence of copper, probably a copper-oxygen complex, induce piperidine-labile damage especially at thymines in the 5′-GTC-3′ sequences. UVA-induced mutations were observed at thymines in rodent and human cells [2, 20]. Although it is known that UV radiation induces the formation of thymine dimers, UVA-induced DNA photodamage at thymines would be mainly due to the formation of oxidative products of thymine.

Although UVB has been believed to be responsible for solar carcinogenesis, UVA-induced DNA damage in the presence of endogenous photosensitizers may play an important role in solar carcinogenesis. In addition, certain drugs, such as quinolone antibacterials, are capable of causing DNA damage in a similar manner, and such drugs can act as exogenous photosensitizers.

References

1 IARC Working Group: IARC Monographs on the Evaluation of the Carcinogenic Risk of Chemicals to Humans. Lyon, IARC, 1992, vol 55.

2 Drobetsky EA, Turcotte J, Châteauneuf A: A role for ultraviolet A in solar carcinogenesis. Proc Natl Acad Sci USA 1995;92:2350–2354.

3 Ito K, Inoue S, Yamamoto K, Kawanishi S: 8-Hydroxydeoxyguanosine formation at the 5′ site of 5′-GG-3′ sequences in double-stranded DNA by UV radiation with riboflavin. J Biol Chem 1993; 268:13221–13227.

4 Ito K, Kawanishi S: Photoinduced hydroxylation of deoxyguanosine in DNA by pterins: Sequence specificity and mechanism. Biochemistry 1997;36:1774–1781.

5 Ito K, Kawanishi S: Site-specific DNA damage induced by UVA radiation in the presence of endogenous photosensitizers. Biol Chem 1997;378:1307–1312.

6 Klecak G, Urbach F, Urwyler H: Fluoroquinolone antibacterials enhance UVA-induced skin tumors. J Photochem Photobiol B Biol 1997;37:174–181.

7 Mäkinen M, Forbes PD, Stenbäck F: Quinolone antibacterials: A new class of photochemical carcinogens. J Photochem Photobiol B Biol 1997;37:182–187.

8 Hiraku Y, Ito H, Kawanishi S: Site-specific hydroxylation at polyguanosine in double-stranded DNA by UVA radiation with nalidixic acid. Biochem Biophys Res Commun 1998;251:466–470.

9 Sugiyama H, Saito I: Theoretical studies of GG-specific photocleavage of DNA via electron transfer: Significant lowering of ionization potential and 5′-localization of HOMO of stacked GG bases in B-form DNA. J Am Chem Soc 1996;118:7063–7068.

10 Yoshioka Y, Kitagawa Y, Takano Y, Yamaguchi K, Nakamura T, Saito I: Experimental and theoretical studies on the selectivity of GGG triplets toward one-electron oxidation in B-form DNA. J Am Chem Soc 1999;121:8712–8719.

11 Burrows CJ, Muller JG: Oxidative nucleobase modifications leading to strand scission. Chem Rev 1998;98:1109–1151.

12 Hall DB, Holmlin RE, Barton JK: Oxidative DNA damage through long-range electron transfer. Nature 1996;382:731–735.

13 Nunez ME, Hall DB, Barton JK: Long-range oxidative to DNA: Effects of distance and sequence. Chem Biol 1999;6:85–97.

14 Kawanishi S, Inoue S, Sano S, Aiba H: Photodynamic guanine modification by hematoporphyrin is specific for single-stranded DNA with singlet oxygen as a mediator. J Biol Chem 1986;261:6090–6095.

15 Domagala JM: Structure-activity and structure-side-effect relationships for the quinolone antibacterials. J Antimicrob Chemother 1994;33:685–706.
16 Celander DW, Cech TR: Iron(II)-ethylenediaminetetraacetic acid catalyzed cleavage of RNA and DNA oligonucleotides: Similar reactivity toward single- and double-stranded forms. Biochemistry 1990;29:1355–1361.
17 Yamamoto K, Kawanishi S: Hydroxyl free radical is not the main active species in site-specific DNA damage induced by copper(II) ion and hydrogen peroxide. J Biol Chem 1989;264:15435–15440.
18 Shibutani S, Takeshita M, Grollman AP: Insertion of specific bases during DNA synthesis past the oxidation-damaged base 8-oxodG. Nature 1991;349:431–434.
19 Ananthaswamy HN, Pierceall WE: Molecular mechanisms of ultraviolet radiation carcinogenesis. Photochem Photobiol 1990;52:1119–1136.
20 Robert C, Muel B, Benoit A, Dubertret L, Sarasin A, Stary A: Cell survival and shuttle vector mutagenesis induced by ultraviolet A and ultraviolet B radiation in a human cell line. J Invest Dermatol 1996;106:721–728.

Shosuke Kawanishi, Department of Hygiene, Mie University School of Medicine,
Tsu, Mie 514-8507 (Japan)
Tel./Fax +81 59 231 5011, E-Mail kawanisi@doc.medic.mie-u.ac.jp

Thiele J, Elsner P (eds): Oxidants and Antioxidants in Cutaneous Biology.
Curr Probl Dermatol. Basel, Karger, 2001, vol 29, pp 83–94

UV-Induced Oxidative Stress and Photoaging

Jutta Wenk, Peter Brenneisen, Christian Meewes, Meinhard Wlaschek, Thorsten Peters, Ralf Blaudschun, Wenijan Ma, Lale Kuhr, Lars Schneider, Karin Scharffetter-Kochanek

Department of Dermatology, University of Cologne, Germany

Aging of the skin has fascinated researchers for decades not only to ultimately prevent wrinkle formation, but also because the skin represents an excellent and accessible model organ allowing the study of intrinsic and extrinsic factors coordinately contributing to the complex phenomenon of aging. Basic principles underlying skin aging are thought to have general relevance for common degenerative connective-tissue diseases like osteoarthritis, osteoporosis and arteriosclerosis [1]. Chronological (intrinsic) aging affects the skin in a manner similar to that of other organs [2]. Superimposed on this innate process, extrinsic aging is related to environmental, mainly UV-induced damage of the dermal connective tissue. There is evidence that these processes, intrinsic and extrinsic aging, have at least in part overlapping, biological, biochemical and molecular mechanisms [3]. Cellular changes as well as qualitative and quantitative alterations of dermal extracellular matrix proteins are involved, resulting in loss of recoil capacity and tensile strength with wrinkle formation, increased fragility and impaired wound healing. UVB (280–320 nm) and UVA (320–400 nm) are essential components of sunlight that generate severe oxidative stress in skin cells via interaction with intracellular chromophores and photosensitizers, resulting in transient and permanent genetic damage, and in the activation of cytoplasmic signal transduction pathways that are related to growth, differentiation, replicative senescence and connective-tissue degradation. In this review, we will discuss photoaging as related to its morphological phenotype and underlying mechanisms.

Clinical and Biochemical Features in Photoaging

Photoaging of the skin is a complex biological process affecting various layers of the skin with the major damage seen in the connective tissue of the dermis. Therefore, in this review, we will concentrate mainly on changes in the connective tissue of the skin. The dermis lies below the epidermis and in conjunction with the basement membrane at the dermal-epidermal junction provides mechanical support for the outer protective layers of the epidermis. Two overlapping, simultaneously occurring processes contribute to the overall changes in skin aging. The first comprises innate or intrinsic aging mechanisms which – similarly to internal organs – affect the skin by a slow and partly reversible degeneration of connective tissue. The second process designated as extrinsic or photoaging is mainly due to UV radiation of sunlight which overwhelmingly contributes to a premature aging phenotype even in young individuals. While the consequences of intrinsic aging can be evaluated in areas protected from the sun, sun-exposed areas like the face and the backside of the hands reveal the overall damage from the innate and the extrinsic aging processes. Clinically, photoaging is characterized by wrinkles, laxity, a leathery appearance, increased fragility, blister formation and impaired wound healing. By contrast, intrinsically aged skin is thin and has reduced elasticity, but is smooth. At the histological level, intrinsically aged skin shows general atrophy of the extracellular matrix with decreased elastin and disintegration of elastic fibers [4]. Photoaged skin is characterized by an increase in deposition of glycosaminoglycans and dystrophic elastotic material in the deep dermis which reveal immunopositive staining for severely disorganized tropoelastin and its associated microfibrillar component fibrillin [5, 6]. By contrast, using immunostaining and confocal microscopy the microfibrillar component fibrillin appeared significantly truncated and depleted in the upper dermis at the dermal-epidermal junction of photoaged skin [7]. In situ hybridization revealed decreased fibrillin 1 mRNA but unchanged fibrillin 2 mRNA levels in severely photodamaged forearm biopsies relative to photoprotected dermal sites. This, in conjunction with the depletion of fibrillin at the dermal-epidermal junction, suggests extensive remodeling at photoaged skin sites [7]. Elastic fibers constitute structural elements of the connective tissue that have a central core of amorphous, hydrophobic cross-linked elastin surrounded by fibrillin-rich microfibrils. As the elastic fiber network extends from the dermal-epidermal junction to the deep dermis, the observed changes may contribute to the clinical features of photoaging such as loss of elasticity.

In addition, photoaged skin is characterized by a loss of mature dermal collagen and histologically by a distinct basophilic appearance of collagen ('basophilic degeneration').

Based on quantitative determination of hydroxyproline by HPLC, Northern blot analysis and immunostaining with antibodies directed against the aminopropeptide of newly synthesized collagen type I, which constitutes the major structural component of the dermal connective tissue, hydroxyproline has been found to be diminished in photoaged skin [8–10].

Collagen type I belongs to a family of closely related but genetically distinct proteins providing the dermis with tensile strength and stability. Also, anchoring fibrils containing collagen type VII which contribute to the stabilization of the epidermal-dermal junction are severely reduced in photoaged skin.

Because the enzymatic capacity for extracellular matrix degradation resides in dermal fibroblasts and in inflammatory cells which are increased in photodamaged skin as well, much effort has been devoted to study the UV-dependent regulation of these degradative processes. Besides UV-affected post-translational modifications of the newly synthesized collagen molecule [3], a variety of laboratories including ours have shown that various matrix metalloproteases, serine and other proteases, responsible for the breakdown of various connective-tissue components were dose-dependently induced in vitro and in vivo by UVA and UVB irradiation [11–18]. The family of matrix metalloproteinases (MMPs) is growing and comprises at least 19 members [19]. While MMP-1 (interstitial collagenase) cleaves collagen type I, MMP-2 is able to degrade elastin as well as basement membrane compounds including collagen type IV and type VII. MMP-3 reveals the broadest substrate specificity for proteins such as collagen type IV, proteoglycans, fibronectin and laminin.

Furthermore, fibrillin has recently been reported to be a target for the proteolytic attack of matrix metalloproteases [20] and MMP-independent serine proteinase activities such as neutrophil elastase. Apart from changes in the organization of the structural components of the connective tissue, also the resident fibroblasts of the dermal connective tissue reveal characteristic features in the photoaged skin. The fibroblast adopts a stellate phenotype with an increase in the rough endoplasmic reticulum indicating biosynthetic activity [2]. Futhermore, an increase in mast cells and neutrophils has been reported in photoaged skin [21, 22].

UV-Generated 'Oxidative Stress' Drives Photoaging

The increase in UV irradiation on earth due to stratospheric ozone depletion represents a major environmental threat to the skin increasing its risk of photooxidative damage by reactive oxygen species (ROS). Though ROS are part of normal regulatory circuits and the cellular redox state is tightly controlled by antioxidants, increased ROS load and loss of cellular redox homeostasis can promote carcinogenesis and photoaging. Loss of cellular redox homeostasis is causally linked to UV-generated ROS substantially compromising the enzymatic and nonenzymatic antioxidant defense of the skin [23, 24], thus tilting the balance towards a prooxidant state [25, 26]. The resulting oxidative stress causes damage to cellular components and changes the pattern of gene expression finally leading to skin pathologies such as nonmelanoma and melanoma skin cancers, phototoxicity and photoaging [18, 27–37]. There is accumulating evidence for the damaging effects of higher concentrations of ROS generated in vitro and in vivo following UVA and UVB irradiation of the skin [38]. Besides direct absorption of UVB photons by DNA and subsequent structural changes, generation of ROS following irradiation with UVA and UVB requires the absorption of photons by endogenous photosensitizer molecules. Recently, the identification of the epidermal UVA-absorbing chromophore trans-urocanic acid that quantitatively accounts for the action spectrum of photoaging has been reported [39]. The excited photosensitizer subsequently reacts with oxygen, resulting in the generation of ROS including the superoxide anion ($O_2^{\cdot -}$) and singlet oxygen ($^1O_2$). $O_2^{\cdot -}$ and $^1O_2$ are also produced by neutrophils that are increased in photodamaged skin and contribute to the overall prooxidant state. Superoxide dismutase converts $O_2^{\cdot -}$ to hydrogen peroxide (H_2O_2). H_2O_2 is able to cross cell membranes easily and in conjunction with transitional Fe(II) drives the generation of the highly toxic hydroxyl radical (HO·) which can initiate lipid peroxidation of cellular membranes [13] with the generation of carbonyls and to date poorly understood consequences.

UV-Induced Reactive Oxygen Species and Their Damaging Effect on Structural Dermal Proteins

Elastin accumulation and collagen degradation are prominent hallmarks in photodamaged skin. Recently, ROS have been reported to enhance tropoelastin mRNA levels [40]. ROS also play a substantial role in collagen metabolism. They not only directly destroy interstitial collagen, but also inactivate tissue inhibitors of metalloproteases and induce the synthesis and activation of matrix-degrading metalloproteases. During the past years, we have used different strategies to study the effect of distinct ROS on the regulation of different matrix-degrading metalloproteases. Exposure of fibroblast monolayer culture to ROS-generating systems or UV irradiation at different spectra in the presence and absence of ROS-quenching/scavenging agents or substances which specifically inhibit ROS-detoxifying enzymes allows to increase or decrease ROS intra- and pericellularly. Also, stably transfected cell lines overexpressing antioxidant enzymes and iron chelators blocking the Fenton reaction have been used. Based on this combined approach, indirect evidence was provided that $^1O_2$ and H_2O_2 are the major ROS involved in the UVA-dependent induction of MMP-1, MMP-2 and MMP-3 on mRNA and protein levels [13, 18, 41–43], while the hydroxyl radical and intermediates of lipid peroxidation play a major role in the UVB induction of MMP-1 and MMP-3 [13]. To mimic an enhanced ROS load, we generated stably transfected fibroblast cell lines with isolated, unbalanced overexpression of manganese superoxide dismutase (MnSOD) [43] (fig. 1a). UVA irradiation of exclusively MnSOD-overexpressing fibroblasts with subsequent intracellular accumulation of hydrogen peroxide enhanced early events in the downstream signaling which resulted in an up to 9.5-fold increase in matrix-degrading MMP-1 mRNA levels compared to vector-transfected control cells (fig. 1b). A similar increase in MMP-1 mRNA was also seen when the intracellular H_2O_2 concentration was increased by the inhibition of different H_2O_2-detoxifying pathways (fig. 1c). Furthermore, UVA irradiation led to a strong induction of *c-jun* and *c-fos* mRNA levels resulting in a 4-fold higher *trans*-activation of the transcription factor AP-1 in the MnSOD-overexpressing cells (fig. 1d). These results perfectly fit an earlier published model for the increased load of ROS in carcinogenesis [44] and photoaging [43] in that imbalances in the interrelated and interdependent antioxidant enzymes drive the accumulation of intracellular ROS, as in our case hydrogen peroxide, which subsequently activates signal transduction pathways and modulates the acitivity of genes that regulate effector genes related to connective-tissue degradation.

Apart from stimulation of signal transduction pathways by intracellularly increased H_2O_2 levels following UVA irradiation [43], there are independent data which provide evidence that UVA-generated singlet oxygen may initiate a membrane-dependent signaling pathway involving c-Jun-N-terminal kinase (JNK) and p38 [59] members of the mitogen-activated protein kinases, and interrelated antocrine cytokine loops IL-1α, IL-1β and IL-6 leading to the enhanced expression of matrix metalloproteases [18, 42, 45]. Interestingly, UVA-generated singlet oxygen has recently been identified to cause the mitochondrial common deletion [46], which had earlier been shown by the same group to occur at higher rates in photoaged skin. This may be of relevance as the common deletion compromises the mitochondrial oxidative phosphorylation, thus further enhancing the overall ROS load.

There is in vitro evidence from fibroblast monolayer cultures that the UVB-initiated iron-driven Fenton reaction with subsequent generation of hydroxyl radicals and lipid peroxidation end products such as malondialdehyde and 4-hydroxy-2(E)-nonenal stimulate the c-Jun amino-terminal kinase 2 representing, beside ERK and p38 kinases, an additional family of the mitogen-

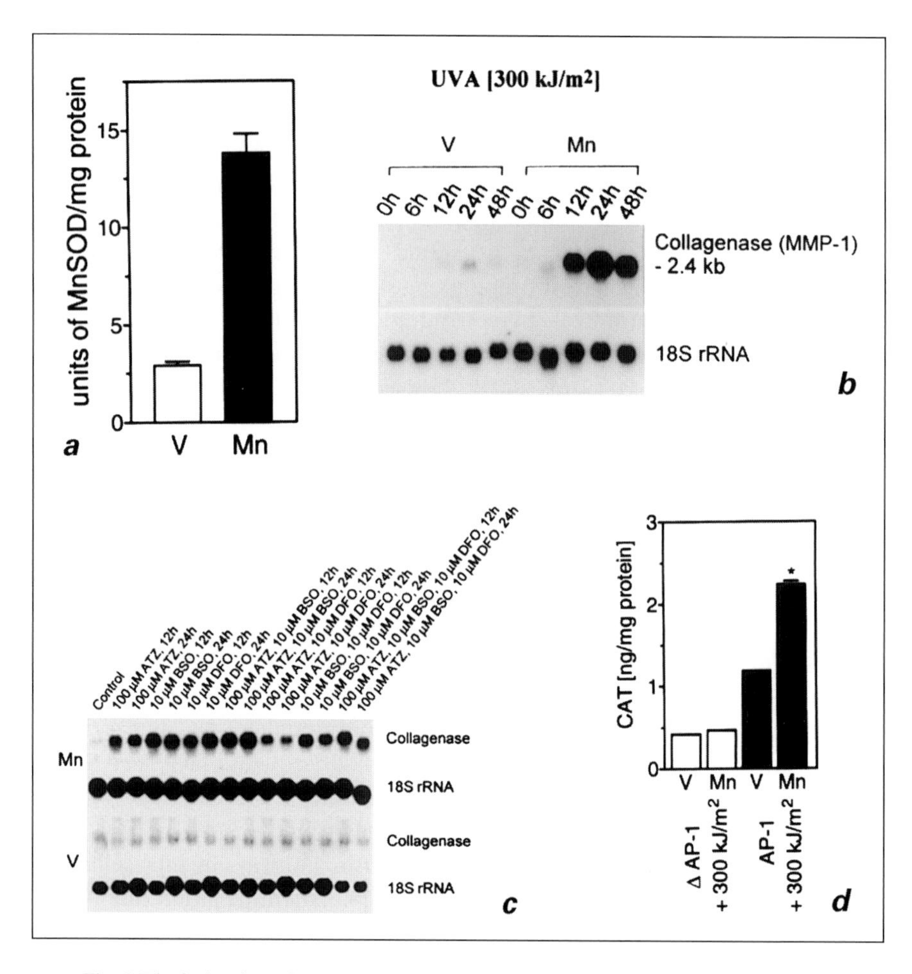

Fig. 1. The induction of MMP-1 mRNA in MnSOD-overexpressing cells upon exposure to UVA irradiation or different prooxidant compounds is mediated via stronger trans-activation of AP-1. ***a*** MnSOD activity in MnSOD-overexpressing cells (Mn) compared to vector control cells (V). ***b*** UVA irradiation resulted in a stronger induction of MMP-1 mRNA in MnSOD-overexpressing cells (Mn) compared to vector control (V). ***c*** Induction of steady-state mRNA levels of MMP-1 in MnSOD-overexpressing and vector-transfected control cells by different prooxidant compounds. MnSOD-overexpressing cells (Mn) and vector-transfected control cells (V) were incubated with ATZ (aminotriazole), an inhibitor of catalase, and BSO (buthionine sulfoximine), an indirect inhibitor of glutathione peroxidase. Both agents led to a further increase in intracellular hydrogen peroxide. Furthermore, MnSOD-overexpressing and vector-transfected cells were incubated with the iron chelator DFO (deferoxamine) for 12 and 24 h to block the H_2O_2-consuming Fenton reaction. Total RNA was isolated and subjected to Northern blot analysis. ***d*** UVA irradiation results in a stronger trans-activation of AP-1 in MnSOD-overexpressing cells compared with vector-transfected control cells. Stably MnSOD-overexpressing cells (Mn) and vector-transfected control

activated protein kinases. UVB-induced c-Jun amino-terminal kinase 2 leads to the phosphorylation and activation of c-Jun protein which upregulates its own expression. Elevated c-Jun, in combination with constitutively expressed c-Fos increases the transcription of MMPs [12, 14]. Recently, we have found that DNA-damage-dependent Fk 506-binding protein 12/rapamycin-associated protein kinase (FRAP) and the p70 ribosomal S6 kinase are critically involved in the UVB induction of MMPs in fibroblast monolayer cultures suggesting that apart from direct DNA damage ROS-induced DNA damage may play a role in the UVB-initiated signal transduction pathway resulting in MMP induction [47]. These data may have considerable in vivo relevance as (1) chronic exposure of hairless mice (Skh-1) and human skin with suberythemal doses substantially increased levels of nonheme iron in the skin and (2) topical application of iron chelators to hairless mouse skin significantly delayed photoaging. Recently, Fisher et al. [14] have shown that upon UV irradiation of human skin mitogen-activated protein kinase signal transduction pathways including c-Jun amino-terminal kinase and p38 resulted in AP-1 activation with subsequently enhanced expression and activation of different MMPs.

Most importantly, these authors provide in vivo evidence that topical treatment of human skin with the vitamin A derivative retinoic acid 48 h prior to UVB irradiation led to the inhibition of MMP expression. Hence, in addition to the well-established capacity of retinoic acid to repair dermal damage of photoaged skin [8, 48, 49], retinoic acid reveals protective properties against photoaging. This is particularly relevant as UV irradiation results in a functional deficiency of vitamin A in the skin [50].

UV-Induced Reactive Oxygen Species and Their Damaging Effects on the Cellular Component of Dermal Connective Tissue

Though controversially discussed, there is evidence that replicative senescence may also occur in vivo with a continuous accumulation of senescent

cells (V) were transfected with plasmids containing a CAT gene either driven by a MMP-1 promoter [–517/+63 TRE Coll CAT (AP-1)] or by an MMP-1 promoter containing a mutated AP-1 site [+63 TRE Coll CAT (ΔAP-1)]. The transfected cells were UVA irradiated and assayed for the amount of synthesized CAT protein by a specific enzyme-linked immunosorbent assay. CAT expression was normalized to β-galactosidase expression. The data represent the mean of 3 independent experiments. SD < 10%, *p = 0.0023 compared with vector-transfected 1306 fibroblasts (V) (Student's t test). From Wenk et al. [43], with permission of the *Journal of Biological Chemistry.*

cells in tissues, where changes of their phenotype like overexpression of matrix-degrading metalloproteases [51] may contribute to age-related pathology of the connective tissue of the skin or other organs as summarized in Campisi [52]. From the notion that ROS are involved in replicative senescence, intrinsic and extrinsic aging may come from several models including models of cell biology, naturally occurring genetic disorders and transgenic organisms. The hypothesis that ROS derived from oxidative metabolism drive the aging process was earlier forwarded by Harman [53] and is based on the observation that about 2% of oxygen taken up is chemically reduced in the mitochondria by the addition of single electrons, which are sequentially converted into ROS. In this context, in vitro experiments with fibroblast cultures showed that α-phenyl-t-butylnitrone, an antioxidant, substantially delays the onset of growth cessation in spontaneously aging fibroblasts. Furthermore, overexpression of antioxidant enzymes detoxifying superoxide anion and hydrogen peroxide or mutations enhancing the ROS-detoxifying enzymatic activities increased the organismic lifespan of aging models like *Drosophila* and *Caenorhabditis elegans* reviewed in Johnson et al. [54], while an imbalance in antioxidant enzymes appears to be an important determinant of cellular senescence. In fact, cells exclusively overexpressing the Cu, Zn superoxide dismutase as a consequence of gene dosage in Down's syndrome (trisomy 21) or due to stable transfection display many features of senescence [55, 56].

The natural shift towards a more prooxidant state in intrinsically aged skin can be significantly enhanced following UV irradiation. However, to the best of our knowledge there are only few reports on UV effects on ROS-induced replicative senescence or the underlying molecular mechanism. Fibroblasts derived from skin biopsies of psoriasis patients which had undergone PUVA therapy (psoralen + UVA) reveal a striking decrease in the cumulative population doubling, a parameter reflecting cellular senescence [57]. We have recently shown that treatment of fibroblast monolayer cultures with 8-methoxypsoralen and subsequent UVA irradiation resulted in a permanent switch of mitotic to postmitotic fibroblasts. A single exposure of fibroblasts to 8-methoxypsoralen and UVA resulted in a 5.8-fold upregulation of two matrix-degrading enzymes (MMP-1 and MMP-3) over an observed period of >120 days, while tissue inhibitor of metalloproteinase 1, the major inhibitor of MMP-1 and MMP-3, was only slightly induced [58]. This imbalance between matrix-degrading metalloproteinases and their inhibitors may lead to connective-tissue damage, a hallmark of premature aging. Superoxide anion and hydrogen peroxide were identified as important intermediates in the downstream signaling pathway leading to these complex fibroblast responses upon psoralen photoactivation [58].

Perspectives

As shown above, ROS play a major role in photoaging and induce changes in gene expression pathways related to collagen degradation and elastin accumulation. There is evidence that singlet oxygen and possibly other ROS as generated by UVA irradiation result in common deletion mutation of mitochondrial DNA and most likely via disruption of the oxidative phosphorylation increase the overall ROS load, subsequently activating the transcription of matrix-metalloprotease-encoding genes. Retinoic acid, known for its efficacy to repair photoaged skin, has now been shown to interfere with the UV/ROS-initiated signal transduction pathways of matrix metalloprotease induction and thus – if topically applied prior to UV irradiation – at least partly to prevent photoaging. A rational design of other antioxidants for topical and systemic treatment depends on our understanding of the molecular mechanism and the identification of distinct oxygen species in photoaging and intrinsic aging. This appears particularly useful, as both photoaging and intrinsic aging partly share underlying pathogenic mechanisms.

References

1 Krieg T, Hein R, Mauch C, Aumailley M: Molecular and clinical aspects of connective tissue. Eur J Clin Invest 1988;18:105–123.
2 Uitto J: Connective tissue biochemistry of the aging dermis: Age related alterations in collagen and elastin. Dermatol Clin 1986;4:433–446.
3 Oikarinen A: The aging of skin: Chronoaging versus photoaging. Photodermatol Photoimmunol Photomed 1990;7:3–4.
4 Braverman IM, Fornferko E: Studies on cutaneous ageing. I. The elastic fiber network. J Invest Dermatol 1982;78:434–443.
5 Mitchell RE: Chronic solar dermatosis: A light and electron microscopic study of the dermis. J Invest Dermatol 1967;43:203–230.
6 Werth VP, Kalathil SE, Jaworsky C: Elastic fiber-associated proteins of skin in development and photoaging. Photochem Photobiol 1996;63:308–313.
7 Watson REB, Griffiths CEM, Craven NM, Shuttleworth A, Kielty CM: Fibrillin-rich microfibrils are reduced in photoaged skin: Distribution at the dermal-epidermal junction. J Invest Dermatol 1999;112:782–787.
8 Griffiths CEM, Russman AN, Majmudar G, Singer RS, Hamilton TA, Voorhees JJ: Restoration of collagen formation in photodamaged human skin by tretinoin (retinoic acid). N Engl J Med 1993;329:530–535.
9 Trautinger F, Mazzucco K, Knobler RM, Trenz A, Kokoschka EM: UVA- and UVB-induced changes in hairless mouse skin collagen. Arch Dermatol Res 1994;286:490–494.
10 Talwar HS, Griffiths CEM, Fisher GJ, Hamilton TA, Voorhees JJ: Reduced type I and type III procollagens in photodamaged adult human skin. J Invest Dermatol 1995;105:285–290.
11 Brenneisen P, Oh J, Wlaschek M, Wenk J, Briviba K, Hommel C, Herrmann G, Sies H, Scharffetter-Kochanek K: UVB-wavelength dependence for the regulation of two major matrix-metalloproteinases and their inhibitor TIMP-1 in human dermal fibroblasts. Photochem Photobiol 1996; 64:649–657.

12 Brenneisen P, Briviba K, Wenk J, Wlaschek M, Scharffetter-Kochanek K: Hydrogen peroxide (H_2O_2) increases the steady-state mRNA levels of collagenase/MMP-1 in human dermal fibroblasts. Free Radic Biol Med 1997;22:515–524.
13 Brenneisen P, Wenk J, Klotz LO, Wlaschek M, Brivibas K, Krieg T, Sies H, Scharffetter-Kochanek: Central role of ferrous/ferric iron in the ultraviolet B irradiation-mediated signaling pathway leading to increased interstitial collagenase (matrix-degrading metalloprotease (MMP)-1) and stromelysin-1 (MMP-3) mRNA levels in cultured human dermal fibroblasts. J Biol Chem 1998;273:5279–5287.
14 Fisher GJ, Talwar HS, Lin J, Lin P, McPhillips F, Wang ZQ, Li X, Wan Y, Kang S, Voorhees JJ: Retinoic acid inhibits induction of c-Jun protein by ultraviolet radiation that occurs subsequent to activation of mitogen-activated protein kinase pathways in human skin in vivo. J Clin Invest 1998; 101:1432–1440.
15 Koivukangas V, Kallioinen M, Autio-Harmainen H, Oikarinen A: UV irradiation induces the expression of gelatinases in human skin in vivo. Acta Derm Venereol 1994;74:279–282.
16 Petersen MJ, Hansen C, Craig S: Ultraviolet A irradiation stimulates collagenase production in cultured human fibroblasts. J Invest Dermatol 1992;99:440–444.
17 Scharffetter K, Wlaschek M, Hogg A, Bolsen K, Schothorst A, Goerz G, Krieg T, Plewig G: UVA irradiation induces collagenase in human dermal fibroblasts in vitro and in vivo. Arch Dermatol Res 1991;283:506–511.
18 Scharffetter-Kochanek K, Wlaschek M, Briviba K, Sies H: Singlet oxygen induces collagenase expression in human skin fibroblasts. FEBS Lett 1993;331:304–306.
19 Stetler-Stevenson WG, Hewitt R, Corcoran M: Matrix metalloproteinases and tumor invasion: From correlation and causality to the clinic. Semin Cancer Biol 1996;7:147–154.
20 Ashworth JL, Murphy M, Rock MJ, Sherratt MJ, Shapiro SD, Shuttleworth CA, Kielty CM: Fibrillin degradation by matrix metalloproteinases: Implications for connective tissue remodelling. Biochem J 1999;340:171–181.
21 Kligman LH, Murphy GF: Ultraviolet B radiation increases hairless mouse mast cells in a dose-dependent manner and alters distribution of UV-induced mast cell growth factor. Photochem Photobiol 1996;63:123–127.
22 Lavker RM, Kligman AM: Chronic heliodermatitis: A morphologic evaluation of chronic actinic dermal damage with emphasis on the role of mast cells. J Invest Dermatol 1988;90:325–330.
23 Biesalski HK, Hemmes C, Hopfenmüller W, Schmid C, Gollnick HP: Effects of controlled exposure of sunlight on plasma and skin levels of β-carotene. Free Radic Res 1996;24:215–224.
24 Witt EH, Motchnik P, Packer L: Evidence for UV light as an oxidative stressor in skin; in Fuchs J, Packer L (eds): Oxidative Stress in Dermatology. New York, Dekker, 1993.
25 Sies H: Biochemistry of oxidative stress. Angew Chem 1986;25:1058–1071.
26 Sies H: Oxidative Stress: Oxidants and Antioxidants. New York, Academic Press, 1991.
27 Epstein JH: Photomedicine; in Smith KC (ed): The Science of Photobiology. New York, Plenum Press, 1989, pp 155–192.
28 Gallagher RP, Elwood JM, Yang CP: Is chronic sunlight exposure important in accounting for increase in melanoma incidence? Int J Cancer 1989;44:813–815.
29 Henriksen T, Dahlback A, Larsen SHH, Moan J: Ultraviolet-radiation and skin cancer: Effect of an ozone layer depletion. Photochem Photobiol 1990;51:579–582.
30 Kligman LH: UVA induced biochemical changes in hairless mouse skin collagen: A contrast to UVB effects; in Urbach F (ed): Biological Responses to Ultraviolet A Radiation. Overland Park, Valdemar Publishing, 1992, pp 209–216.
31 Kraemer KM: Sunlight and skin cancer: Another link revealed. Proc Natl Acad Sci USA 1997;94: 11–14.
32 Oikarinen A, Karvonen J, Uitto J, Hannuksela M: Connective tissue alterations in skin exposed to natural and therapeutic UV-radiation. Photodermatology 1985;2:15–26.
33 Oikarinen A, Kallioinen M: A biochemical and immunohistochemical study of collagen in sun-exposed and protected skin. Photodermatology 1989;6:24–31.
34 Scharffetter-Kochanek K, Wlaschek M, Bolsen K, Herrmann G, Lehmann P, Goerz G, Mauch C, Plewig G: Mechanisms of cutaneous photoaging; in Plewig G, Marks R (eds): The Environmental Threat of the Skin. London, Dunitz, 1992, pp 72–82.

35 Scharffetter-Kochanek K, Krieg T: Lichtalterung; in Plewig G, Korting HC (eds): Fortschritte der praktischen Dermatologie und Venerologie. Berlin, Springer, 1995, vol 14, pp 176–179.

36 Scharffetter-Kochanek K: Photodamage of the skin: Prevention and therapy; in Sies H (ed): Antioxidants in Disease – Mechanisms and Therapeutic Strategies. Adv Pharmacol. San Diego, Academic Press, 1997, pp 639–655.

37 Urbach F: Potential effects of altered solar ultraviolet radiation on human skin cancer. Photochem Photobiol 1989;50:507–514.

38 Jurkiewicz BA, Buettner GR: Ultraviolet-light-induced free radical formation in skin: An electron paramagnetic resonance study. Photochem Photobiol 1994;59:1–4.

39 Hanson KM, Simon JD: Epidermal trans-urocanic acid and the UV-A-induced photoaging of the skin. Proc Natl Acad Sci USA 1998;95:10576–10578.

40 Kawaguchi Y, Tanaka H, Okada T, Konishi H, Takahashi M, Ito M, Asai M: Effect of reactive oxygen species on the elastin mRNA expression in cultured human dermal fibroblasts. Free Radic Biol Med 1997;23:162–165.

41 Herrmann G, Wlaschek M, Bolsen K, Prenzel K, Goerz G, Scharffetter-Kochanek K: Pathogenic implication of matrix-metalloproteinases (MMPs) and their counteracting inhibitor TIMP-1 in the cutaneous photodamage of human porphyria cutanea tarda (PCT). J Invest Dermatol 1996;107: 398–403.

42 Wlaschek M, Briviba K, Stricklin GP, Sies H, Scharffetter-Kochanek K: Singlet oxygen may mediate the ultraviolet A-induced synthesis of interstitial collagenase. J Invest Dermatol 1995;104:194–198.

43 Wenk J, Brenneisen P, Wlaschek M, Poswig A, Briviba K, Oberley TD, Scharffetter-Kochanek K: Stable overexpression of manganese superoxide dismutase in mitochondria identifies hydrogen peroxide as a major oxidant in the AP-1-mediated induction of matrix-degrading metalloproteinase-1. J Biol Chem 1999;274:25869–25876.

44 Cerutti PA: Oxy-radicals and cancer. Lancet 1994;344:862–863.

45 Wlaschek M, Wenk J, Brenneisen P, Briviba K, Schwarz A, Sies H, Scharffetter-Kochanek K: Singlet oxygen is an early intermediate in cytokine-dependent ultraviolet-A induction of interstitial collagenase in human dermal fibroblasts in vitro. FEBS Lett 1997;413:239–242.

46 Berneburg M, Grether-Beck S, Kürten V, Ruzicka T, Briviba K, Sies H, Krutmann J: Singlet oxygen mediates the UVA-induced generation of the photoaging-associated mitochondrial common deletion. J Biol Chem 1999;274:15345–15349.

47 Brenneisen P, Wenk J, Wlaschek M, Krieg T, Scharffetter-Kochanek K: Activation of p70 ribosomal protein S6 kinase is an essential step in the DNA damage-dependent signaling pathway responsible for the ultraviolet B-mediated increase in interstitial collagenase (MMP-1) and stromelysin-1 (MMP-3) protein levels in human dermal fibroblasts. J Biol Chem 2000;275:4336–4344.

48 Kligman AM, Grove GL, Hirose R: Topical tretinoin for photoaged skin. J Am Acad Dermatol 1986;15:836–859.

49 Ellis CN, Weiss JS, Hamilton TA, Headington JT, Zelickson AS, Voorhees JJ: Sustained improvement with prolonged topical tretinoin (retinoic acid) for photoaged skin. J Am Acad Dermatol 1990;23: 629–637.

50 Wang Z, Boudjelal M, Kang S, Voorhees JJ, Fisher GJ: Ultraviolet irradiation of human skin causes functional vitamin A deficiency, preventable by all-*trans* retinoic acid pre-treatment. Nat Med 1999; 5:418–422.

51 West MD, Pereira-Smith OM, Smith JR: Replicative senescence of human skin fibroblasts correlates with a loss of regulation and overexpression of collagenase activity. Exp Cell Res 1989;184:138–147.

52 Campisi J: Replicative senescence: An old lives tale? Cell 1996;84:497–500.

53 Harman D: Aging: A theory based on free radical and radiation chemistry. J Gerontol 1956;11: 298–300.

54 Johnson FB, Sinclair DA, Guarente L: Molecular biology of aging. Cell 1999;96:291–302.

55 Ceballos J, Nicole A, Braind P, Grimbler G, Delacoute A, Flament S, Blouin JL, Thevenin M, Klamoun P, Sinet PM: Expression of human Cu-Zn superoxide dismutase gene in transgenic mice: Model for gene dosage effect in Down syndrome. Free Radic Res Commun 1991;12–13:581–589.

56 de Haan JB, Christiano F, Iannello RC, Bladier C, Kelner MJ, Kola I: Elevation in the ratio of Cu/Zn superoxide dismutase to glutathione peroxidase activity induces features of cellular senescence and this effect is mediated by hydrogen peroxide. Hum Mol Genet 1996;5:283–292.
57 Jacobson-Kram D, Roe JL, Williams JR, Gange RW, Parrish JA: Decreased in vitro lifespan of fibroblasts derived from skin exposed to photochemotherapy in vivo. Lancet 1982;18:1399–1400.
58 Herrmann G, Brenneisen P, Wlaschek M, Wenk J, Faisst K, Quel G, Hommel C, Goerz G, Ruzicka T, Krieg T, Sies H: Psoralen photoactivation promotes morphological and functional changes in fibroblasts in vitro reminiscent of cellular senescence. J Cell Sci 1998;111:759–767.
59 Klotz LO, Pellieuse C, Briviba K, Pielot C, Aubry JM, Sies H: Mitogen-activated protein kinase (p38-, JNK-, ERK-) activation pattern induced by extracellular and intracellular singlet oxygen and UVA. Eur J Biochem 1999;260:917–922.

Karin Scharffetter-Kochanek, MD, Department of Dermatology,
Joseph-Stelzmann-Strasse 9, D–50935 Cologne (Germany)
Tel. +49 221 478 6591, Fax +49 221 478 6438, E-Mail Karin.Scharffetter@uni-koeln.de

Thiele J, Elsner P (eds): Oxidants and Antioxidants in Cutaneous Biology.
Curr Probl Dermatol. Basel, Karger, 2001, vol 29, pp 95–113

UVA and Singlet Oxygen as Inducers of Cutaneous Signaling Events

Lars-Oliver Klotz[a], *Nikki J. Holbrook*[b], *Helmut Sies*[a]

[a] Institut für Physiologische Chemie I, Heinrich-Heine-Universität Düsseldorf, Germany;
[b] Laboratory of Biological Chemistry, National Institute on Aging, NIH, Baltimore, Md., USA

UV radiation has traditionally been divided into defined regions of wavelengths, including UVA (320–400 nm), UVB (280–320 nm) and UVC (at wavelengths below 280 nm). Physiologically, the latter is of least importance, since the stratospheric ozone layer hinders UVC from reaching the earth's surface. Nevertheless, it has been successfully employed for establishing biological responses to UV stress [1–4]. Of the three regions, UVA is the one penetrating the skin most deeply, reaching subepidermal layers [5].

UVA has been shown to play a role not only in the context of damage to biomolecules and cells [6, 7], but also to be capable of inducing more complex responses such as the induction of genes. Many of these processes have been demonstrated to be mediated by singlet oxygen [for a review, see 8].

Singlet oxygen is an electronically excited form of molecular oxygen that exists in two forms in solution (fig. 1a), $^1\Delta_g\ O_2$ and $^1\Sigma_g^+\ O_2$; the latter is generally regarded to be too short-lived to be of any significance in biological systems; for a recent review on 'singlet sigma', see Weldon et al. [9]. The term 'singlet oxygen' will therefore refer to the $^1\Delta_g$ form in this article.

This chapter will review the modes of generation of singlet oxygen in biological systems and the role of singlet oxygen in mediating signaling processes induced by UVA and by photodynamic therapy (PDT). Finally, it will be attempted to relate these data to what is known about the effects of UVA and PDT as well as $^1O_2$ on cell survival and apoptosis.

Generation of Singlet Oxygen in vivo

Singlet oxygen may be generated in vivo either photochemically or metabolically – i.e. in dark reactions. The former is limited to body areas exposed to light and is based upon the photodynamic principle of absorption of light by photosensitizing molecules that are capable of transferring photoenergy to triplet ground state molecular oxygen (present in micromolar concentrations in biological systems), thereby elevating it into the excited singlet state (fig. 1b). An example of such photosensitizers are porphyrins. The photosensitizing effect of porphyrins can be observed clinically in certain porphyrias. These disorders in heme biosynthesis are characterized by elevated levels of photosensitizing heme precursor porphyrins or side products and are often associated with increased cutaneous photosensitivity. Porphyrins are photosensitizers not only upon irradiation with visible light, but also in the UVA region [10]. Other endogenous sensitizers that may lead to singlet oxygen production upon irradiation with UVA are flavins and certain quinones [11]. This principle of a sensitized generation of singlet oxygen and other reactive species is exploited medically in photodynamic therapy where different sensitizers are administered and allowed to accumulate in target areas prior to irradiation in order to kill the desired tissue [12, 13].

Metabolic generation of singlet oxygen has been shown to occur in stimulated neutrophils [14]. Singlet oxygen seems to be produced in a series of reactions involving myeloperoxidase, which utilizes hydrogen peroxide and chloride to form hypochlorite, which, in turn, may form singlet oxygen in a reaction with hydrogen peroxide (fig. 1c). From experiments with stimulated macrophages that express phagocytic NADPH oxidase, a multisubunit enzyme-generating superoxide from oxygen at the expense of NADPH [15], but which lacks myeloperoxidase, it was deduced that additional $^1O_2$ may be derived from the spontaneous rather than enzymatic decomposition of superoxide [16]. These data lend support to the much debated theory of $^1O_2$ generation during spontaneous (but not the enzymatically catalyzed) dismutation of superoxide, which implies that superoxide dismutase indirectly protects from $^1O_2$ [17, 18].

Singlet oxygen has been found to be generated in other reactions of biological relevance, such as lipid peroxidation [for a review, see 19, 20] (fig. 1d), the reaction of hydroperoxides with peroxynitrite [21, 22] or of hydrogen peroxide with hypohalites [23]. Further, $^1O_2$ seems to play a role in other enzymatic reactions, such as those catalyzed by prostaglandin hydroperoxidase [24] or the cytochrome P450 complex [25].

The proof for the generation of singlet oxygen in a biological system is based on its physical and chemical properties. The transition of $^1\Delta_g$ singlet

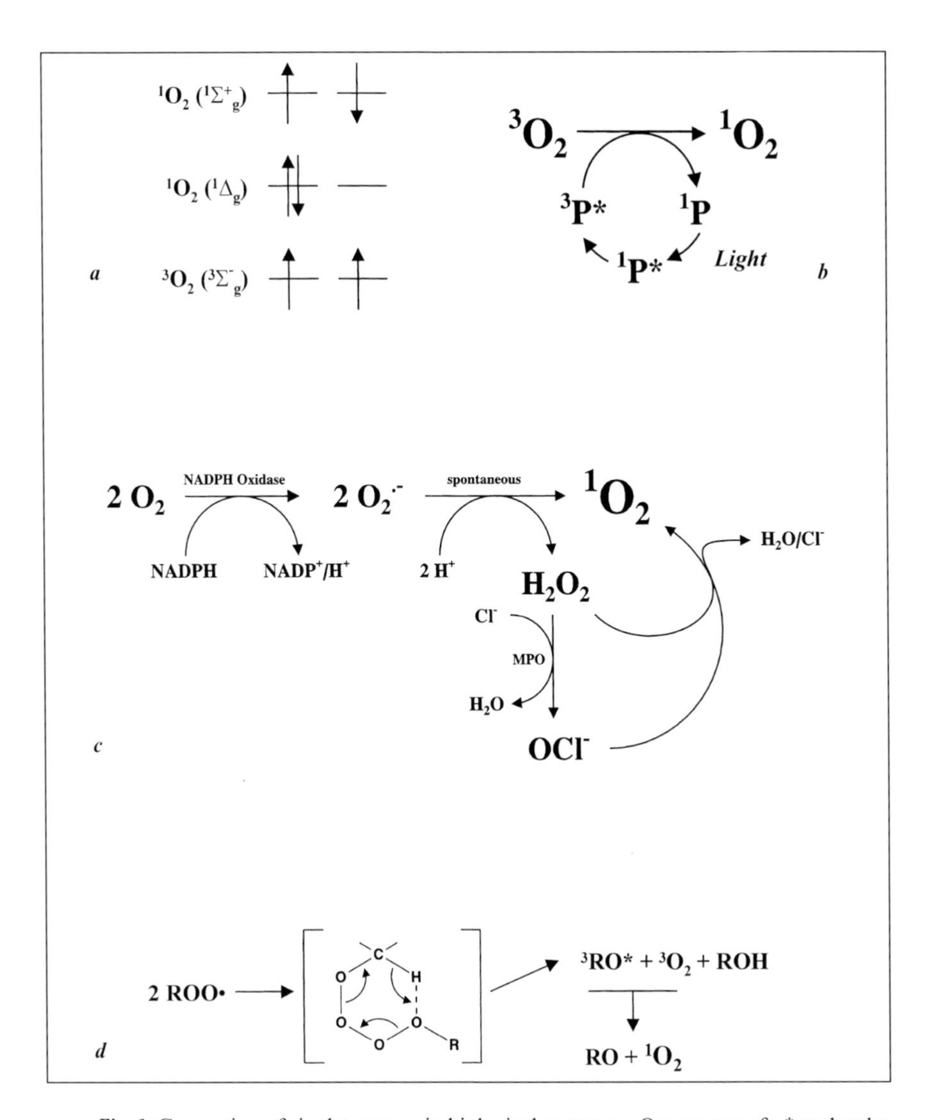

Fig. 1. Generation of singlet oxygen in biological systems. ***a*** Occupancy of π^* molecular orbitals of dioxygen in its triplet ground state and its two excited singlet states. ***b*** Simplified scheme depicting the photochemical generation of singlet oxygen: a photosensitizer in its (singlet) ground state (^1P) is elevated into an excited singlet state ($^1P^*$) by the absorption of light of appropriate wavelength. Following nonradiative decay (intersystem crossing) to the excited triplet state ($^3P^*$), the sensitizer may now react with triplet ground state molecular oxygen ($^3O_2$) to form singlet oxygen ($^1O_2$). Due to spin conservation rules, P must regain a singlet state. ***c*** Sequence of reactions leading to the formation of singlet oxygen in stimulated neutrophils – see text for details. ***d*** Generation of singlet oxygen during lipid peroxidation via formation of triplet excited carbonyls ($^3RO^*$): Russel mechanism of combination of two peroxyl radicals.

oxygen to ground state triplet oxygen is accompanied by emissions in the visible [634 nm; dimol emission ($2^1\Delta_gO_2 \rightarrow 2^3\Sigma_g^-O_2 + h\nu$; 2×94.3 kJ/mol)] and infrared [1,268 nm; monomol emission ($^1\Delta_gO_2 \rightarrow {}^3\Sigma_g^-O_2 + h\nu$; 94.3 kJ/mol)] regions. Whereas the infrared emission at 1,268 nm is regarded as specific for singlet oxygen, chemiluminescence in the 634-nm spectral region is a more common effect and needs further verification to be identified as due to $^1O_2$. Due to the sensitivity of the equipment employed for these measurements, emissions of $^1O_2$ generated under the influence of light are difficult to identify. In such cases, modulation of $^1O_2$ lifetime helps establish the presence of $^1O_2$ and its role in biological effects.

The lifetime of singlet oxygen may be decreased by the addition of quenchers such as carotenoids or azide, or scavengers such as tocopherols, histidine/imidazole and furan derivatives. All of these compounds are also frequently used in cell culture. Despite the known side effects of azide (e.g. as an inhibitor of heme proteins), it is useful in short-term experiments as it quickly enters the cell and may thus be employed without extensive preincubation.

In biological systems, the most active $^1O_2$ quenchers are carotenoids [27]. Due to their lipophilic nature, it is not easy to load cells in culture with defined amounts of carotenoids, which will also tend to accumulate in lipophilic regions of a cell. Further, it seems that under some circumstances the effects of carotenoids cannot be ascribed solely to their $^1O_2$-quenching properties. For example, although $^1O_2$ has been shown to mediate the induction of heme oxygenase 1 (HO-1) by UVA, which is inhibitable by histidine or azide, β-carotene apparently enhances the induction of HO-1 by UVA [28]. This would certainly have led to contradictory results had β-carotene been used as a quencher in the original work showing the involvement of $^1O_2$ in the induction of HO-1 by UVA [29].

While the above agents decrease $^1O_2$ lifetime by either physically or chemically interacting with it, the lifetime and steady-state concentration of $^1O_2$ may also be enhanced by changing the solvent. Singlet oxygen lifetime varies with the solvent, being in the range of 4 μs in H_2O and up to >1 ms in solvents lacking hydrogen and deuterium atoms, such as CCl_4 [30]. In biological systems, there is only one alternative solvent to water, deuterium oxide (D_2O), which enhances $^1O_2$ lifetime about 10-fold. Hence, incubation in the presence of D_2O should increase the steady-state concentration of $^1O_2$ and intensify $^1O_2$-dependent effects.

Finally, singlet oxygen traps, compounds that react with $^1O_2$ to yield specific and more or less stable reaction products that can be analyzed by HPLC or other techniques, can be used to assess the production of $^1O_2$. Examples of such traps are certain anthracene derivatives [14, 16, 31] or cholesterol [32, 33], which may be regarded as an endogenous trap.

Singlet oxygen is chemically more reactive than ground state (triplet) dioxygen and reacts with all major classes of biomolecules, leading to protein oxidation, lipid peroxidation and DNA damage [8, 30 and references therein]. While this reactivity might suggest that the effects of $^1O_2$ on cells are primarily deleterious, it has been shown recently that nontoxic or mildly toxic doses lead to the activation of cellular signaling pathways culminating in modulated gene expression. Moreover, $^1O_2$ mediates the induction of some signaling pathways by UVA.

Singlet Oxygen as Mediator of UVA-Induced Signaling

Our knowledge about signaling caused by $^1O_2$ or UVA is rather fragmentary. No complete pathway – from initial activation of a signaling molecule by $^1O_2$ or UVA to the final expression of a certain gene – has been demonstrated so far. However, educated guesses of the signaling patterns can be made from the current fragmentary knowledge. Two proteins will be focused on here, HO-1 and interstitial collagenase (matrix metalloproteinase 1, MMP-1), products of the first genes shown to be induced by $^1O_2$ and UVA via $^1O_2$ [29, 34, 35]. The induction of these genes by UVA can be simulated with $^1O_2$-generating systems, inhibited by quenchers such as sodium azide and histidine, and intensified in the presence of D_2O, giving strong evidence not only for the formation of $^1O_2$ by UVA, but also for the importance of this intermediate formation in the induction process.

MMP-1 can be quite easily assigned a role in photoaging and photocarcinogenesis: the enhanced degradation of extracellular matrix will favor wrinkle formation, as well as metastasis [36]. The role of HO-1 is less obvious. HO-1 is a microsomal enzyme that catalyzes the rate-limiting step in heme catabolism, i.e. the oxygen- and NADPH-consuming breakdown of heme to form carbon monoxide, ferrous iron and biliverdin which, in turn, is reduced to bilirubin by biliverdin reductase. Therefore, increased HO-1 activity should elevate cellular levels of reactive iron, a process that may be regarded as prooxidative. On the other hand, its induction was proposed to be a beneficial adaptive response to oxidative stress [37, 38] as synthesis of the iron storage protein ferritin is induced upon UVA treatment, most likely as a response to the increases in iron release due to enhanced HO-1 activity [39]. Moreover, both biliverdin and bilirubin exhibit antioxidant properties [40]. The overall singlet-oxygen-quenching constant (chemical plus physical quenching) for biliverdin and bilirubin was determined as 2.3 and 3.2×10^9 $M^{-1}s^{-1}$, respectively [41].

It has been demonstrated that UVA is capable of inducing the activity of the transcription factor activating protein 1 (AP-1) [42, 43] and expression of its component proteins c-Fos [43, 44] and c-Jun [43] in human skin fibroblasts. Upstream of AP-1 are the extracellular signal-regulated (ERK), c-Jun N-terminal (JNK) and p38 mitogen-activated protein (MAP) kinases that, once activated, lead to activation of AP-1 via their phosphorylation of transcription factors such as c-Jun, ATF-2 or Elk-1, and the subsequent induced expression of *c-jun* or *c-fos* [45–47]. As a consequence of activation of AP-1, genes with functional AP-1 responsive elements in their promoter regions (such as HO-1 [48, 49] and MMP-1 [50]) may be induced. Indeed, the p38 and JNK MAP kinases were shown to be activated by UVA and $^1O_2$ [51]; while it is known that this activation is transient [51], the identity of the 'off switch' is not yet known. One candidate is MAP kinase phosphatase 1 (MKP-1), which is induced by UVA [52]. Originally termed CL100, MKP-1 is a dual specificity (Tyr- and Ser/Thr) phosphatase, capable of dephosphorylating and inactivating p38 [53] and JNK MAP kinases [53, 54]. In addition, it is known that MKP-1 can be induced via p38 [55] or JNK [56] by stimuli other than UVA. The scenario might therefore be as follows: UVA leads to the generation of $^1O_2$ via photosensitization. Singlet oxygen then activates JNK and p38 followed by recruitment of AP-1 leading to the induction of HO-1, MMP-1 and MKP-1. MKP-1 may in turn serve as the terminating signal (fig. 2). The mechanism of activation of p38 and JNK by $^1O_2$ is yet to be determined.

It is almost certainly not that simple. First, there is no information on whether authentic singlet oxygen induces AP-1 or leads to the induction of MKP-1. Moreover, the induction of MMP-1 by UVA/$^1O_2$ was found to be mediated by a mechanism different from the above hypothesis (fig. 2): IL-1 and IL-6 are sequentially released upon UVA/$^1O_2$ treatment and, in turn, lead to induction of MMP-1 transcription [57, 58]. Also, it seems that immediate activation of p38 and JNK does not play an important role in the induction of MMP-1 or HO-1: p38 and JNK activation by $^1O_2$ was shown to be caused by intracellularly formed singlet oxygen, generated by rose bengal + light or the lipophilic naphthalene derivative $DHPNO_2$, but not by $^1O_2$ formed extracellularly with agarose-coupled rose bengal or the hydrophilic $^1O_2$ generator $NDPO_2$ [51]; both of these extracellular $^1O_2$ sources, however, are capable of inducing MMP-1 and HO-1 [35, 58, Klotz et al., unpubl.]. Further, there are other pathways induced by both UVA and $^1O_2$, such as nuclear factor (NF) κB [59, 60] – and functional response elements for this transcription factor are present in both HO-1 [48] and IL-6 [61] promoter regions (fig. 2). Yet NF-κB, like p38 and JNK, seems to be activatable by intracellularly generated $^1O_2$ [60], but not by an extracellular generator like $NDPO_2$ [62]. Hence, NF-κB is probably not a major contributor to induction of MMP-1, IL-6 or HO-1 by $^1O_2$.

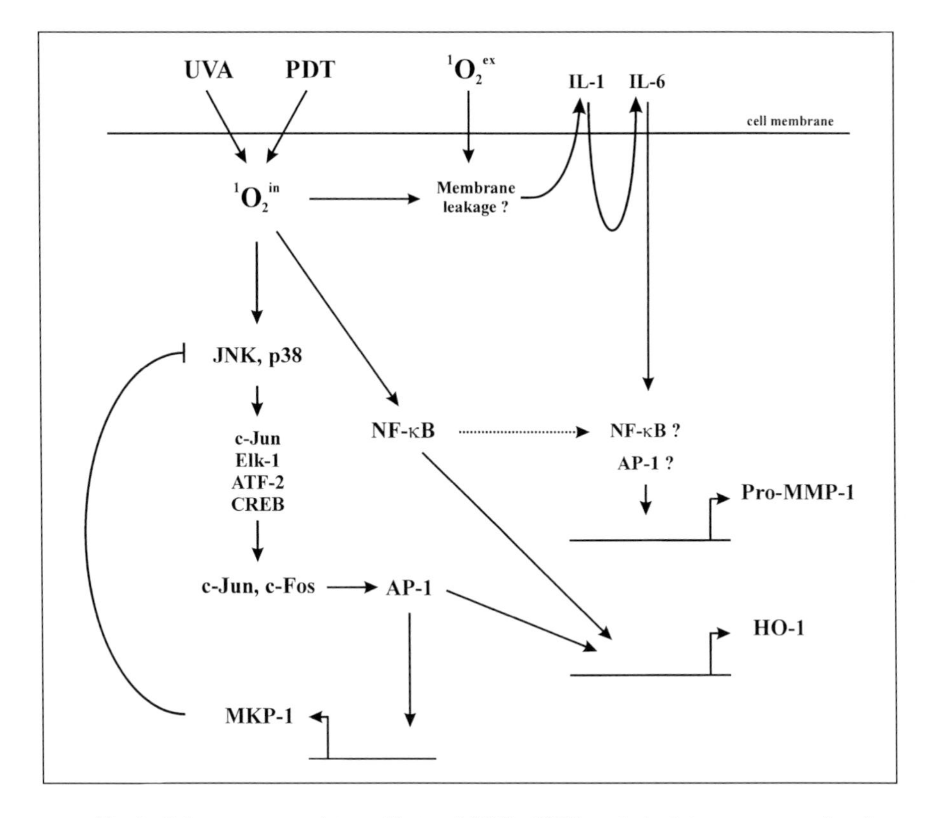

Fig. 2. Scheme summarizing effects of UVA, PDT and singlet oxygen on signaling pathways leading to modulated gene expression. The scheme combines well-documented phenomena with mostly hypothetical pathways. UVA and PDT lead to the intracellular generation of singlet oxygen ($^1O_2^{in}$) which was shown to activate JNK and p38 MAP kinases and NF-κB. Both may contribute to the induction of MMP-1 and HO-1 along the shown pathways. Both MMP-1 and HO-1, but neither JNK, p38 nor NF-κB, are activated by extracellularly generated singlet oxygen ($^1O_2^{ex}$). The induction of MMP-1 has been shown to be mediated by autocrine loops with release of IL-1 and IL-1-induced release of IL-6 from the cell. Singlet-oxygen-induced membrane leakage has been made responsible for the release of preformed IL-1 from the cell. A role of AP-1 or NF-κB, both of which are activated by IL-6, in the $^1O_2$-induced induction of MMP-1 or HO-1, is still hypothetical. See text for more detail.

A possible site where $^1O_2$ initiates the processes leading to MMP-1 and HO-1 induction is the cell membrane or some molecule closely associated with it and thereby accessible to extracellularly formed $^1O_2$. Basu-Modak et al. [7] showed that lipid peroxidation caused by UVA is probably not responsible for the induction of HO-1 per se, as lipophilic antioxidants such as butylated hydroxytoluene or α-tocopherol did not inhibit the induction, though they suppressed lipid peroxidation. The cyclooxygenase inhibitor indomethacin, however, did suppress induction of HO-1 by UVA. In agreement with this, phospholipase A_2 has been shown to be activated upon UVA irradiation, as is cyclooxygenase [63, 64]. Activation of cyclooxygenase apparently leads to a release of heme from microsomal hemeproteins [65]; heme, in turn, is a strong activator of HO-1.

Another UVA-induced signaling pathway leading to the induction of intercellular adhesion molecule 1 (ICAM-1) via $^1O_2$ has been identified in human keratinocytes [66, 67]. The transcription factor AP-2 has been shown to mediate this process. The question of what is upstream of AP-2 and downstream of $^1O_2$ is not yet completely resolved, but there is evidence that formation of ceramides upon treatment might be close to the top of the cascade [68]. Further, this process is most likely connected with the cell membrane, which might explain that it is accessible to $^1O_2$ derived from $NDPO_2$, which can also stimulate ICAM-1 expression.

In summary, three pathways have been shown to be induced by $^1O_2$ and/or UVA, the AP-1, NF-κB and AP-2 pathways. Further, the cell membrane appears to play a role as one of the primary targets for $^1O_2$ and UVA leading to the enhanced expression of MMP-1, HO-1 and ICAM-1. On the other hand, there must be additional intracellular targets as the p38/JNK and NF-κB pathways are activatable by intracellularly formed $^1O_2$ only.

Signaling Induced by Photodynamic Therapy

Singlet oxygen is generated photochemically in PDT. Hence, it is no surprise that many of the above effects have been shown for treatments of cells with various combinations of photosensitizers + light. Singlet oxygen is assumed to be the reactive species responsible for the observed biological effects, but in most studies a causal relation to $^1O_2$ is not firmly established, as other reactive oxygen species (e.g. formed in type I photochemical processes [8]) have not been ruled out. That can be done by either employing the above-mentioned strategies of modulating $^1O_2$ lifetime or by a priori defining conditions under which it is highly unlikely that any reactive oxygen species other than $^1O_2$ is formed [69, 70].

This section will briefly discuss PDT employing visible light and photosensitizers of the porphyrin and phthalocyanine family. PUVA (5-methoxypsoralen + UVA) therapy will not be discussed.

Many of the genes induced by PDT are also inducible by UVA (table 1) including HO-1, certain interleukins and others. Regarding upstream events

Table 1. Products of genes induced by UVA, PDT or singlet oxygen

	UVA		PDT		$^1O_2$	
Bax	+	[93]				
Bcl-2	↓	[92, 93]				
c-Fos	+	[43, 44]	+	[74, 80]	+	[80]
c-Jun	+	[43]	+	[74]		
Clusterin	+	[113]				
c-Myc	(+)	[114]	+	[80]		
Collagen VII	+	[115]				
Cox-2	+	[116]				
Egr-1	+	[117]	+	[80]		
FasL	+	[91]			+	[91]
Fra-1	+	[114]				
GM-CSF	+	[118]				
Grp-78			+	[82]		
HO-1	+	[37]	+	[119]	+	[29]
Hsp-70	+	[120]	+	[81]		
ICAM-1	+	[66]			+	[66]
IL-1α/β	+	[57]			+	[58]
IL-6	+	[57]	+	[75, 121]	+	[58]
IL-8	+	[122]				
IL-10	+	[122]	+	[121]		
MKP-1 (=CL100)	+	[52]				
MMP-1	+	[123]			+	[34, 35]
MMP-2	+	[123]				
MMP-3	+	[123]				
p21			+	[124]		
p53	+	[104, 105]				
TNF-α	+	[122]				

+ = Induction; ↓ = downregulation; Bax = Bcl-2-associated X protein; Bcl = B-cell leukemia/lymphoma; Cox = cyclooxygenase; Egr = early growth response gene; Fas = CD95/APo-1; Fra = Fos-related antigen; Grp = glucose-regulated protein; GM-CSF = granulocyte-macrophage colony-stimulating factor; Hsp = heat shock protein; TNF = tumor necrosis factor.

leading to these inductions, the same holds true for various forms of PDT that were found for UVA. For example, JNK and p38 MAP kinases [71–73], as well as their downstream targets *c-jun* and *c-fos* [74] were demonstrated to be induced, leading to AP-1-dependent gene expression (e.g. of IL-6) [74, 75]. ERK MAP kinases do not seem to be activated significantly by either UVA [51], PDT [71–73] or by singlet oxygen [51, 76]. In parallel, PDT induces

NF-κB [77–79], which seems to play an important role in sensitivity of treated cells in terms of apoptosis (see below).

As with UVA, the cell membrane seems to be of importance, as induction of c-*fos* after photosensitization is inhibited by addition of the phospholipase A_2 inhibitor quinacrine, pointing to similarities with UVA-induced activation of phospholipase A_2 and the concomitant release of arachidonate and arachidonate metabolites [80]. Further, heat shock protein 70 is induced after increased binding of heat shock factor to heat shock response elements [81] and seems to contribute to enhanced tolerance to heat stress following PDT.

Amongst the first genes shown to be induced by PDT was glucose-regulated protein 78 (grp-78) [82]. Grp-78 is a molecular chaperone of 78 kD located in the endoplasmic reticulum (ER). It binds to malfolded proteins and is induced as a result of the 'unfolded protein response'. Thus, grp-78 induction is regarded as a marker for 'ER stress', a term comprising stressful stimuli that lead to defective processing of proteins in the ER [for a recent review, see 83]. Such stimuli are glucose deprivation, inhibitors of protein glycosylation, modulators of calcium homeostasis, but also oxidative stresses that impair the redox environment needed for proper protein folding. Depending on the intracellular localization of the respective sensitizer at the time of irradiation, PDT can easily be imagined to have the same effect. Specifically, PDT could impair the integrity of the ER membrane by inducing lipid peroxidation, thus leading to disruption of calcium homeostasis; it could also oxidize proteins, thus preventing their proper processing. PDT, and probably singlet oxygen, may thus be regarded as ER stresses, though certainly their effects are not restricted or specific to the ER.

The following section will briefly discuss how the above-mentioned effects of UVA, PDT and singlet oxygen relate to and can be integrated into processes determining whether a cell exposed to stress will survive or die.

UVA, PDT, Singlet Oxygen and Apoptosis

It is a matter of viewpoint whether or not the killing of cells under the influence of all singlet-oxygen-generating conditions is desirable, but certainly this is the case for the therapeutic application of UVA and PDT. In UVA phototherapy, mostly employing the UVA1 (340–400 nm) region of the spectrum for the treatment of cutaneous disorders such as atopic dermatitis [84] or systemic lupus erythematosus [85], and in PDT, which is applied to eliminate hyperproliferative cells from an otherwise healthy context [13], the generation of intermediate reactive species such as singlet oxygen helps induce cell death. At high-dose treatment death occurs mainly via necrosis, while at lower doses apoptosis predominates. The former is the result of overwhelming incident stress, while the latter involves the cell's systematic self-destruction, without affecting the surrounding tissue.

Godar [86, 87] differentiates three kinetic forms of apoptosis after treatment of cells with UV of different wavelengths: (1) immediate apoptosis, due to direct targeting of mitochondria by the respective stress with subsequent release of apoptosis-inducing factor or cytochrome c, is evident within half an hour after insult; (2) intermediate apoptosis, involving receptor-triggered mechanisms such as FasR/FasL interactions leading to the activation of initiator caspases like caspase 8, commences within 4 h of treatment, and (3) delayed apoptosis, or genuine 'programmed cell death' requiring the synthesis of proteins after stimulus, is induced later than 4 h after stress. DNA damage is believed to be the trigger of events for this late onset of cell death.

Induction of apoptosis by UVA has been shown in a variety of cellular systems including murine lymphoma cells [88], rat [89] and human [90] fibroblasts, human T [86, 91] and B [86] lymphocytes. The induction of apoptosis is immediate and leads to a downregulation of the mitochondrial antiapoptotic protein Bcl-2 [92, 93] and an upregulation of the proapoptotic Bax [93]. In agreement with these data, overexpression of Bcl-2 [89], or induction of Bcl-2 by treatment with nitrogen monoxide [93], was demonstrated to inhibit UVA-induced immediate apoptosis. As the induction of the oxidative stress-responsive HO-1 was obstructed in the former case, the authors proposed that Bcl-2 might act via an antioxidant pathway [89]. The lipophilic antioxidant α-tocopherol was also shown to interrupt UVA-induced immediate apoptosis, suggesting that membrane damage is an important starting point in the process [88]. The target may well be mitochondrial membranes, since mitochondrial transmembrane potential is disrupted, and cyclosporin A, a blocker of the redox-sensitive 'S site' of the mitochondrial permeability transition pore [94], decelerates the apoptotic process [86]. Immediate apoptosis of T and B lymphocytes after UVA could be influenced by modulators of singlet oxygen steady-state concentrations; sodium azide (but not cyanide) inhibited and D_2O enhanced the process [86, 91]. Moreover, both the $^1O_2$ generator $NDPO_2$ as well as irradiation of cells treated with rose bengal mimicked the apoptotic events including an upregulation of FasL by $^1O_2$ or UVA (via $^1O_2$). These data suggest that, in addition to the immediate and direct mitochondrial processes, the Fas/FasL system might play a role in the rapid induction of apoptosis [91]. Release of cytochrome c from mitochondria during immediate and intermediate apoptosis leads to activation of caspases 9 and 3, followed by cellular disintegration at the protein level [95]. Both cytochrome c release and caspase 3 activation have been shown for treatments with UVA [96], singlet oxygen and various forms of PDT [79, 97–100], implying the activation of caspase 9. In the case of singlet oxygen, human promyelocytic leukemia cells (HL-60) were irradiated with white light in the presence of rose bengal [97]; activation of caspase 3 was mediated by caspase 8, as was cytochrome c release. Inhibition of either caspase 3 or caspase 8 blocked the apoptotic process in these cells. The authors discuss a role of caspase-8-mediated release of cytochrome c, probably via cleavage of Bid, a mediator of Fas-induced apoptosis [101], in augmenting the signal of caspase 8 that results in cleavage and activation of procaspase 3. The same $^1O_2$-generating system induces apoptosis in HeLa cells and in primary human skin fibroblasts [Klotz et al., unpubl.]. Caspase 8 can also be activated by Fas ligation [102], which, as mentioned above, has indeed been shown to occur after treatment of T lymphocytes with UVA or $^1O_2$ [91]. There are, however, reports on Fas-independent activation of caspase 8 [103].

Since not only UVC and UVB, but also UVA and PDT lead to a considerable degree of DNA damage in irradiated cells [6, 96], a role of p53 in UVA-induced apoptosis of the delayed kind can be envisaged. Indeed, UVA irradiation of human skin leads to induced levels of p53, predominantly in the basal epidermal layers [104, 105]. However, many cell types fail to show elevation of p53 levels in response to UVA [92], and the role of p53 in PDT is still controversial with reports ranging from increased sensitivity of cells harboring wild-type p53 towards PDT [106–108] to no effect of p53 disruption on sensitivity of cells at all [109].

UVA, PDT and $^1O_2$ were all shown to activate the transcription factor NF-κB (see above) that has been ascribed antiapoptotic properties, which could indeed be demonstrated for PDT of human colon cancer cells [79]. Activation of NF-κB by PDT is likely to proceed via formation of ceramides [78].

UVA [51], various forms of PDT [71–73] and $^1O_2$ [51] have all been shown to activate a specific set of MAP kinase family members: these treatments activate JNK and p38 MAP kinases, but so far no activation of ERK MAP kinase has been reported. In 1995, Xia et al. [110] proposed a crucial role of p38 and JNK as proapoptotic stimuli in PC12 cells, whereas activation of ERK seemed to be antiapoptotic. Although the importance of JNK and p38 in mediating apoptosis following stress has been well documented in a variety of model systems, there are clear exceptions to this generality, and, in fact, cases in which JNK and/or p38 activations appear to play a role in protecting cells against apoptosis [111, 112]. One example of particular relevance to this review is the case of hypericin-PDT-induced apoptosis [112], where inhibition of JNK and p38 activation, achieved either by use of pharmacological inhibitors or by overexpression of either dominant-negative upstream regulators or dominant-positive MKP-1, sensitized HeLa cells towards PDT. However, in another system [73], the opposite seems to be true: p38 and JNK are activated in murine leukemic lymphoblasts upon PDT employing the phthalocyanine Pc4, and pharmacological inhibition of p38 inhibited apoptosis. In summary, the role of MAP kinases in UVA-/PDT-induced apoptosis is enigmatic and subject to cell type specificity or photosensitizer selectivity.

Conclusion

Singlet oxygen is an important intermediate in the effects of UVA and PDT on cutaneous processes. Knowledge concerning the various pathways induced in human skin by these stimuli and the cellular consequences of their activation is still patchy and incomplete. Further understanding of the mechanisms involved in regulating the cellular response to such treatments will help explain the effectiveness of UVA1 and PDT treatments for cutaneous disorders and likely lead to the development of improved therapeutic strategies. It should also provide better understanding of the mechanisms of UVA-induced photoaging and photocarcinogenesis.

Acknowledgements

L.O.K. is a recipient of a postdoctoral fellowship of the Deutsche Forschungsgemeinschaft, Bonn, Germany (KL 1245/1-1). H.S. is an NFCR fellow. This work was supported by the DFG (SFB503/B1).

References

1 Stein B, Rahmsdorf HJ, Steffen A, Litfin M, Herrlich P: UV-induced DNA damage is an intermediate step in UV-induced expression of human immunodeficiency virus type 1, collagenase, *c-fos*, and metallothionein. Mol Cell Biol 1989;9:5169–5181.
2 Sachsenmaier C, Radler-Pohl A, Zinck R, Nordheim A, Herrlich P, Rahmsdorf HJ: Involvement of growth factor receptors in the mammalian UVC response. Cell 1994;78:963–972.
3 Derijard B, Hibi M, Wu IH, Barrett T, Su B, Deng T, Karin M, Davis RJ: JNK1: A protein kinase stimulated by UV light and Ha-Ras that binds and phosphorylates the c-Jun activation domain. Cell 1994;76:1025–1037.
4 Bender K, Blattner C, Knebel A, Iordanov M, Herrlich P, Rahmsdorf HJ: UV-induced signal transduction. J Photochem Photobiol B 1997;37:1–17.
5 Tyrrell RM: Activation of mammalian gene expression by the UV component of sunlight – From models to reality. Bioessays 1996;18:139–148.
6 Kvam E, Tyrrell RM: Induction of oxidative DNA base damage in human skin cells by UV and near visible radiation. Carcinogenesis 1997;18:2379–2384.
7 Basu-Modak S, Lüscher P, Tyrrell RM: Lipid metabolite involvement in the activation of the human heme oxygenase-1 gene. Free Radic Biol Med 1996;20:887–897.
8 Briviba K, Klotz LO, Sies H: Toxic and signaling effects of photochemically or chemically generated singlet oxygen in biological systems. Biol Chem 1997;378:1259–1265.

9 Weldon D, Poulsen TD, Mikkelsen KV, Ogilby P: Singlet sigma: The 'other' singlet oxygen in solution. Photochem Photobiol 1999;70:369–379.
10 Ryter SW, Tyrrell RM: Singlet molecular oxygen ($^1O_2$): A possible effector of eukaryotic gene expression. Free Radic Biol Med 1998;24:1520–1534.
11 Tyrrell RM: UVA (320–380 nm) radiation as an oxidative stress; in Sies H (ed): Oxidative Stress: Oxidants and Antioxidants. San Diego, Academic Press, 1991, pp 57–83.
12 Peng Q, Berg K, Moan J, Kongshaug M, Nesland JM: 5-Aminolevulinic acid-based photodynamic therapy: Principles and experimental research. Photochem Photobiol 1997;65:235–251.
13 Dougherty TJ, Gomer CJ, Henderson BW, Jori G, Kessel D, Korbelik M, Moan J, Peng Q: Photodynamic therapy. J Natl Cancer Inst 1999;90:889–905.
14 Steinbeck MJ, Khan AU, Karnovsky MJ: Intracellular singlet oxygen generation by phagocytosing neutrophils in response to particles coated with a chemical trap. J Biol Chem 1992;267:13425–13433.
15 Babior BM: NADPH oxidase: An update. Blood 1999;93:1464–1476.
16 Steinbeck MJ, Khan AU, Karnovsky MJ: Extracellular production of singlet oxygen by stimulated macrophages quantified using 9,10-diphenylanthracene and perylene in a polystyrene film. J Biol Chem 1993;268:15649–15654.
17 Corey EJ, Mehrotra MM, Khan AU: Water induced dismutation of superoxide anion generates singlet molecular oxygen. Biochem Biophys Res Commun 1987;145:842–846.
18 Khan AU, Kasha M: Singlet molecular oxygen in the Haber-Weiss reaction. Proc Natl Acad Sci USA 1994;91:12365–12367.
19 Cadenas E: Biochemistry of oxygen toxicity. Annu Rev Biochem 1989;58:79–110.
20 Cadenas E, Sies H: The lag phase. Free Radic Res 1998;28:601–609.
21 Di Mascio P, Bechara EJ, Medeiros MH, Briviba K, Sies H: Singlet molecular oxygen production in the reaction of peroxynitrite with hydrogen peroxide. FEBS Lett 1994;355:287–289.
22 Di Mascio P, Briviba K, Sasaki ST, Catalani LH, Medeiros MH, Bechara EJ, Sies H: The reaction of peroxynitrite with tert-butyl hydroperoxide produces singlet molecular oxygen. Biol Chem 1997;378:1071–1074.
23 Kanofsky JR: Singlet oxygen production by biological systems. Chem Biol Interact 1989;70:1–28.
24 Cadenas E, Sies H, Nastainczyk W, Ullrich V: Singlet oxygen formation detected by low-level chemiluminescence during enzymatic reduction of prostaglandin G_2 to H_2. Hoppe-Seylers Z Physiol Chem 1983;364:519–528.
25 Osada M, Ogura Y, Yasui H, Sakurai H: Involvement of singlet oxygen in cytochrome P450-dependent substrate oxidations. Biochem Biophys Res Commun 1999;263:392–397.
26 Cadenas E, Sies H: Low-level chemiluminescence as an indicator of singlet molecular oxygen in biological systems. Methods Enzymol 1984;105:221–231.
27 Sies H, Stahl W: Vitamins E and C, beta-carotene, and other carotenoids as antioxidants. Am J Clin Nutr 1995;62:1315S–1321S.
28 Obermüller-Jevic UC, Francz PI, Frank J, Flaccus A, Biesalski HK: Enhancement of the UVA induction of haem oxygenase-1 expression by beta-carotene in human skin fibroblasts. FEBS Lett 1999;460:212–216.
29 Basu-Modak S, Tyrrell RM: Singlet oxygen: A primary effector in the ultraviolet A/near-visible light induction of the human heme oxygenase gene. Cancer Res 1993;53:4505–4510.
30 Foote CS, Clennan EL: Properties and reactions of singlet oxygen; in Foote CS, Valentine JS, Greenberg A, Liebman JF (eds): Active Oxygen in Chemistry. London, Blackie Academic, 1995, pp 105–141.
31 Di Mascio P, Sies H: Quantification of singlet oxygen generated by thermolysis of 3,3′-(1,4-naphthylidene)dipropionate: Monomol and dimol photoemission and the effects of 1,4-diazabicyclo[2.2.2.]octane. J Am Chem Soc 1989;111:2909–2914.
32 Girotti AW: Photosensitized oxidation of cholesterol in biological systems: Reaction pathways, cytotoxic effects and defense mechanisms. J Photochem Photobiol B 1992;13:105–118.
33 Korytowski W, Girotti AW: Singlet oxygen adducts of cholesterol: Photogeneration and reductive turnover in membrane systems. Photochem Photobiol 1999;70:484–489.
34 Scharffetter-Kochanek K, Wlaschek M, Briviba K, Sies H: Singlet oxygen induces collagenase expression in human skin fibroblasts. FEBS Lett 1993;331:304–306.

35 Wlaschek M, Briviba K, Stricklin GP, Sies H, Scharffetter-Kochanek K: Singlet oxygen may mediate the ultraviolet A-induced synthesis of interstitial collagenase. J Invest Dermatol 1995;104:194–198.
36 Scharffetter-Kochanek K, Wlaschek M, Brenneisen P, Schauen M, Blaudschun M, Wenk J: UV-induced reactive oxygen species in photocarcinogenesis and photoaging. Biol Chem 1997;378:1247–1257.
37 Keyse SM, Tyrrell RM: Heme oxygenase is the major 32–kDa stress protein induced in human skin fibroblasts by UVA radiation, hydrogen peroxide, and sodium arsenite. Proc Natl Acad Sci USA 1989;86:99–103.
38 Applegate LA, Luscher P, Tyrrell RM: Induction of heme oxygenase: A general response to oxidant stress in cultured mammalian cells. Cancer Res 1991;51:974–978.
39 Vile GF, Tyrrell RM: Oxidative stress resulting from ultraviolet A irradiation of human skin fibroblasts leads to a heme oxygenase-dependent increase in ferritin. J Biol Chem 1993;268:14678–14681.
40 Stocker R, Yamamoto Y, McDonagh AF, Glazer AN, Ames BN: Bilirubin is an antioxidant of possible physiologic importance. Science 1987;235:1043–1046.
41 Di Mascio P, Kaiser S, Sies H: Lycopene as the most efficient biological carotenoid singlet oxygen quencher. Arch Biochem Biophys 1989;274:532–538.
42 Djavaheri-Mergny M, Mergny JL, Bertrand F, Santus R, Mazière C, Dubertret L, Mazière JC: Ultraviolet-A induces activation of AP-1 in cultured human keratinocytes. FEBS Lett 1996;384:92–96.
43 Wenk J, Brenneisen P, Wlaschek M, Poswig A, Briviba K, Oberley TD, Scharffetter-Kochanek K: Stable overexpression of manganese superoxide dismutase in mitochondria identifies hydrogen peroxide as a major oxidant in the AP-1-mediated induction of matrix-degrading metalloprotease-1. J Biol Chem 1999;274:25869–25876.
44 Bose B, Soriani M, Tyrrell RM: Activation of expression of the c-*fos* oncogene by UVA irradiation in cultured human skin fibroblasts. Photochem Photobiol 1999;69:489–493.
45 Karin M: The regulation of AP-1 activity by mitogen-activated protein kinases. J Biol Chem 1995;270:16483–16486.
46 Karin M, Liu Zg, Zandi E: AP-1 function and regulation. Curr Opin Cell Biol 1997;9:240–246.
47 Iordanov M, Bender K, Ade T, Schmid W, Sachsenmaier C, Engel K, Gaestel M, Rahmsdorf HJ, Herrlich P: CREB is activated by UVC through a p38/HOG-1-dependent protein kinase. EMBO J 1997;16:1009–1022.
48 Lavrovsky Y, Schwartzman ML, Levere RD, Kappas A, Abraham NG: Identification of binding sites for transcription factors NF-kappa B and AP-2 in the promoter region of the human heme oxygenase 1 gene. Proc Natl Acad Sci USA 1994;91:5987–5991.
49 Elbirt KK, Whitmarsh AJ, Davis RJ, Bonkovsky HL: Mechanism of sodium arsenite-mediated induction of heme oxygenase-1 in hepatoma cells: Role of mitogen-activated protein kinases. J Biol Chem 1998;273:8922–8931.
50 Angel P, Imagawa M, Chiu R, Stein B, Imbra RJ, Rahmsdorf HJ, Jonat C, Herrlich P, Karin M: Phorbol ester-inducible genes contain a common *cis* element recognized by a TPA-modulated *trans*-acting factor. Cell 1987;49:729–739.
51 Klotz LO, Pellieux C, Briviba K, Pierlot C, Aubry JM, Sies H: Mitogen-activated protein kinase (p38-, JNK-, ERK-) activation pattern induced by extracellular and intracellular singlet oxygen and UVA. Eur J Biochem 1999;260:917–922.
52 Keyse SM, Emslie EA: Oxidative stress and heat shock induce a human gene encoding a protein-tyrosine phosphatase. Nature 1992;359:644–647.
53 Chu Y, Solski PA, Khosravi Far R, Der CJ, Kelly K: The mitogen-activated protein kinase phosphatases PAC1, MKP-1, and MKP-2 have unique substrate specificities and reduced activity in vivo toward the ERK2 sevenmaker mutation. J Biol Chem 1996;271:6497–6501.
54 Liu Y, Gorospe M, Yang C, Holbrook NJ: Role of mitogen-activated protein kinase phosphatase during the cellular response to genotoxic stress: Inhibition of c-Jun N-terminal kinase activity and AP-1-dependent gene activation. J Biol Chem 1995;270:8377–8380.
55 Schliess F, Heinrich S, Haussinger D: Hyperosmotic induction of the mitogen-activated protein kinase phosphatase MKP-1 in H4IIE rat hepatoma cells. Arch Biochem Biophys 1998;351:35–40.

56 Bokemeyer D, Sorokin A, Yan M, Ahn NG, Templeton DJ, Dunn MJ: Induction of mitogen-activated protein kinase phosphatase 1 by the stress-activated protein kinase signaling pathway but not by extracellular signal-regulated kinase in fibroblasts. J Biol Chem 1996;271:639–642.
57 Wlaschek M, Heinen G, Poswig A, Schwarz A, Krieg T, Scharffetter-Kochanek K: UVA-induced autocrine stimulation of fibroblast-derived collagenase/MMP-1 by interrelated loops of interleukin-1 and interleukin-6. Photochem Photobiol 1994;59:550–556.
58 Wlaschek M, Wenk J, Brenneisen P, Briviba K, Schwarz A, Sies H, Scharffetter-Kochanek K: Singlet oxygen is an early intermediate in cytokine-dependent ultraviolet-A induction of interstitial collagenase in human dermal fibroblasts in vitro. FEBS Lett 1997;413:239–242.
59 Vile GF, Tanew Ilitschew A, Tyrrell RM: Activation of NF-kappa B in human skin fibroblasts by the oxidative stress generated by UVA radiation. Photochem Photobiol 1995;62:463–468.
60 Piret B, Legrand Poels S, Sappey C, Piette J: NF-kappa B transcription factor and human immunodeficiency virus type 1 (HIV-1) activation by methylene blue photosensitization. Eur J Biochem 1995;228:447–455.
61 Matsui H, Ihara Y, Fujio Y, Kunisada K, Akira S, Kishimoto T, Yamauchi-Takihara K: Induction of interleukin (IL)-6 by hypoxia is mediated by nuclear factor (NF)-kappa B and NF-IL6 in cardiac myocytes. Cardiovasc Res 1999;42:104–112.
62 Schreck R, Albermann K, Baeuerle PA: Nuclear factor kappa B: An oxidative stress-responsive transcription factor of eukaryotic cells (a review). Free Radic Res Commun 1992;17:221–237.
63 Hanson DL, DeLeo VA: Long wave ultraviolet radiation stimulates arachidonic acid release and cyclooxygenase activity in mammalian cells in culture. Photochem Photobiol 1989;49:423–430.
64 Hanson D, DeLeo V: Long-wave ultraviolet light induces phospholipase activation in cultured human epidermal keratinocytes. J Invest Dermatol 1990;95:158–163.
65 Kvam E, Noel A, Basu-Modak S, Tyrrell RM: Cyclooxygenase dependent release of heme from microsomal hemeproteins correlates with induction of heme oxygenase 1 transcription in human fibroblasts. Free Radic Biol Med 1999;26:511–517.
66 Grether-Beck S, Olaizola-Horn S, Schmitt H, Grewe M, Jahnke A, Johnson JP, Briviba K, Sies H, Krutmann J: Activation of transcription factor AP-2 mediates UVA radiation- and singlet oxygen-induced expression of the human intercellular adhesion molecule 1 gene. Proc Natl Acad Sci USA 1996;93:14586–14591.
67 Grether-Beck S, Buettner R, Krutmann J: Ultraviolet A radiation-induced expression of human genes: Molecular and photobiological mechanisms. Biol Chem 1997;378:1231–1236.
68 Grether Beck S, Schmitt H, Felsner I, Bonizzi G, Piette J, Johnson JP, Klotz LO, Krutmann J: Ceramide signaling is a key component of ultraviolet A radiation-induced gene expression in human keratinocytes. J Invest Dermatol 1998;110:624.
69 Allen MT, Lynch M, Lagos A, Redmond RW, Kochevar IE: A wavelength dependent mechanism for rose bengal-sensitized photoinhibition of red cell acetylcholinesterase. Biochim Biophys Acta 1991;1075:42–49.
70 Lambert CR, Kochevar IE: Does rose bengal triplet generate superoxide anion? J Am Chem Soc 1996;118:3297–3298.
71 Tao J, Sanghera JS, Pelech SL, Wong G, Levy JG: Stimulation of stress-activated protein kinase and p38 HOG1 kinase in murine keratinocytes following photodynamic therapy with benzoporphyrin derivative. J Biol Chem 1996;271:27107–27115.
72 Klotz LO, Fritsch C, Briviba K, Tsacmacidis N, Schliess F, Sies H: Activation of JNK and p38 but not ERK MAP kinases in human skin cells by 5-aminolevulinate-photodynamic therapy. Cancer Res 1998;58:4297–4300.
73 Xue LY, He J, Oleinick NL: Promotion of photodynamic therapy-induced apoptosis by stress kinases. Cell Death Differ 1999;6:855–864.
74 Kick G, Messer G, Plewig G, Kind P, Goetz AE: Strong and prolonged induction of *c-jun* and *c-fos* proto-oncogenes by photodynamic therapy. Br J Cancer 1996;74:30–36.
75 Kick G, Messer G, Goetz A, Plewig G, Kind P: Photodynamic therapy induces expression of interleukin 6 by activation of AP-1 but not NF-kappa B DNA binding. Cancer Res 1995;55:2373–2379.

76 Zhuang S, Lynch MC, Kochevar IE: Activation of protein kinase C is required for protection of cells against apoptosis induced by singlet oxygen. FEBS Lett 1998;437:158–162.
77 Ryter SW, Gomer CJ: Nuclear factor kappa B binding activity in mouse L1210 cells following photofrin II-mediated photosensitization. Photochem Photobiol 1993;58:753–756.
78 Matroule JY, Bonizzi G, Morliere P, Paillous N, Santus R, Bours V, Piette J: Pyropheophorbide-a methyl ester-mediated photosensitization activates transcription factor NF-kappaB through the interleukin-1 receptor-dependent signaling pathway. J Biol Chem 1999;274:2988–3000.
79 Matroule JY, Hellin AC, Morliere P, Fabiano AS, Santus R, Merville MP, Piette J: Role of nuclear factor-kappa B in colon cancer cell apoptosis mediated by aminopyropheophorbide photosensitization. Photochem Photobiol 1999;70:540–548.
80 Luna MC, Wong S, Gomer CJ: Photodynamic therapy mediated induction of early response genes. Cancer Res 1994;54:1374–1380.
81 Gomer CJ, Ryter SW, Ferrario A, Rucker N, Wong S, Fisher AM: Photodynamic therapy-mediated oxidative stress can induce expression of heat shock proteins. Cancer Res 1996;56:2355–2360.
82 Gomer CJ, Ferrario A, Rucker N, Wong S, Lee AS: Glucose regulated protein induction and cellular resistance to oxidative stress mediated by porphyrin photosensitization. Cancer Res 1991; 51:6574–6579.
83 Pahl HL: Signal transduction from the endoplasmic reticulum to the cell nucleus. Physiol Rev 1999; 79:683–701.
84 Krutmann J, Morita A: Mechanisms of ultraviolet (UV) B and UVA phototherapy. J Invest Dermatol Symp Proc 1999;4:70–72.
85 McGrath H Jr: Ultraviolet A1 (340–400 nm) irradiation and systemic lupus erythematosus. J Invest Dermatol Symp Proc 1999;4:79–84.
86 Godar DE: UVA1 radiation triggers two different final apoptotic pathways. J Invest Dermatol 1999; 112:3–12.
87 Godar DE: Light and death: Photons and apoptosis. J Invest Dermatol Symp Proc 1999;4: 17–23.
88 Godar DE, Lucas AD: Spectral dependence of UV-induced immediate and delayed apoptosis: The role of membrane and DNA damage. Photochem Photobiol 1995;62:108–113.
89 Pourzand C, Rossier G, Reelfs O, Borner C, Tyrrell RM: Overexpression of Bcl-2 inhibits UVA-mediated immediate apoptosis in rat 6 fibroblasts: Evidence for the involvement of Bcl-2 as an antioxidant. Cancer Res 1997;57:1405–1411.
90 Leccia MT, Richard MJ, Favier A, Beani JC: Zinc protects against ultraviolet A1-induced DNA damage and apoptosis in cultured human fibroblasts. Biol Trace Elem Res 1999;69:177–190.
91 Morita A, Werfel T, Stege H, Ahrens C, Karmann K, Grewe M, Grether-Beck S, Ruzicka T, Kapp A, Klotz LO, Sies H, Krutmann J: Evidence that singlet oxygen-induced human T helper cell apoptosis is the basic mechanism of ultraviolet-A radiation phototherapy. J Exp Med 1997;186: 1763–1768.
92 Wang Y, Rosenstein B, Goldwyn S, Zhang X, Lebwohl M, Wei H: Differential regulation of P53 and Bcl-2 expression by ultraviolet A and B. J Invest Dermatol 1998;111:380–384.
93 Suschek CV, Krischel V, Bruch-Gerharz D, Berendji D, Krutmann J, Kroncke KD, Kolb-Bachofen V: Nitric oxide fully protects against UVA-induced apoptosis in tight correlation with Bcl-2 up-regulation. J Biol Chem 1999;274:6130–6137.
94 Costantini P, Chernyak BV, Petronilli V, Bernardi P: Modulation of the mitochondrial permeability transition pore by pyridine nucleotides and dithiol oxidation at two separate sites. J Biol Chem 1996;271:6746–6751.
95 Thornberry NA, Lazebnik Y: Caspases: Enemies within. Science 1998;281:1312–1316.
96 Tada-Oikawa S, Oikawa S, Kawanishi S: Role of ultraviolet A-induced oxidative DNA damage in apoptosis via loss of mitochondrial membrane potential and caspase-3 activation. Biochem Biophys Res Commun 1998;247:693–696.
97 Zhuang S, Lynch MC, Kochevar IE: Caspase-8 mediates caspase-3 activation and cytochrome c release during singlet oxygen-induced apoptosis of HL-60 cells. Exp Cell Res 1999;250:203–212.
98 Kim HR, Luo Y, Li G, Kessel D: Enhanced apoptotic response to photodynamic therapy after bcl-2 transfection. Cancer Res 1999;59:3429–3432.

99 Granville DJ, Carthy CM, Jiang H, Shore GC, McManus BM, Hunt DW: Rapid cytochrome c release, activation of caspases 3, 6, 7 and 8 followed by Bap31 cleavage in HeLa cells treated with photodynamic therapy. FEBS Lett 1998;437:5–10.

100 Varnes ME, Chiu SM, Xue LY, Oleinick NL: Photodynamic therapy-induced apoptosis in lymphoma cells: Translocation of cytochrome c causes inhibition of respiration as well as caspase activation. Biochem Biophys Res Commun 1999;255:673–679.

101 Li H, Zhu H, Xu CJ, Yuan J: Cleavage of BID by caspase 8 mediates the mitochondrial damage in the Fas pathway of apoptosis. Cell 1998;94:491–501.

102 Ashkenazi A, Dixit VM: Apoptosis control by death and decoy receptors. Curr Opin Cell Biol 1999;11:255–260.

103 Belka C, Heinrich V, Marini P, Faltin H, Schulze-Osthoff K, Bamberg M, Budach W: Ionizing radiation and the activation of caspase-8 in highly apoptosis-sensitive lymphoma cells. Int J Radiat Biol 1999;75:1257–1264.

104 Campbell C, Quinn AG, Angus B, Farr PM, Rees JL: Wavelength specific patterns of p53 induction in human skin following exposure to UV radiation. Cancer Res 1993;53:2697–2699.

105 Burren R, Scaletta C, Frenk E, Panizzon RG, Applegate LA: Sunlight and carcinogenesis: Expression of p53 and pyrimidine dimers in human skin following UVA I, UVA I+II and solar simulating radiations. Int J Cancer 1998;76:201–206.

106 Fisher AM, Danenberg K, Banerjee D, Bertino JR, Danenberg P, Gomer CJ: Increased photosensitivity in HL60 cells expressing wild-type p53. Photochem Photobiol 1997;66:265–270.

107 Fisher AM, Rucker N, Wong S, Gomer CJ: Differential photosensitivity in wild-type and mutant p53 human colon carcinoma cell lines. J Photochem Photobiol B 1998;42:104–107.

108 Zhang WG, Li XW, Ma LP, Wang SW, Yang HY, Zhang ZY: Wild-type p53 protein potentiates phototoxicity of 2-BA-2-DMHA in HT29 cells expressing endogenous mutant p53. Cancer Lett 1999;138:189–195.

109 Fisher AM, Ferrario A, Rucker N, Zhang S, Gomer CJ: Photodynamic therapy sensitivity is not altered in human tumor cells after abrogation of p53 function. Cancer Res 1999;59:331–335.

110 Xia Z, Dickens M, Raingeaud J, Davis RJ, Greenberg ME: Opposing effects of ERK and JNK-p38 MAP kinases on apoptosis. Science 1995;270:1326–1331.

111 Roulston A, Reinhard C, Amiri P, Williams LT: Early activation of c-Jun N-terminal kinase and p38 kinase regulate cell survival in response to tumor necrosis factor α. J Biol Chem 1998;273:10232–10239.

112 Assefa Z, Vantieghem A, Declercq W, Vandenabeele P, Vandenheede JR, Merlevede W, de Witte P, Agostinis P: The activation of the c-Jun N-terminal kinase and p38 mitogen-activated protein kinase signaling pathways protects HeLa cells from apoptosis following photodynamic therapy with hypericin. J Biol Chem 1999;274:8788–8796.

113 Viard I, Wehrli P, Jornot L, Bullani R, Vechietti JL, Schifferli JA, Tschopp J, French LE: Clusterin gene expression mediates resistance to apoptotic cell death induced by heat shock and oxidative stress. J Invest Dermatol 1999;112:290–296.

114 Ariizumi K, Bergstresser PR, Takashima A: Wavelength-specific induction of immediate early genes by ultraviolet radiation. J Dermatol Sci 1996;12:147–155.

115 Chen M, Petersen MJ, Li HL, Cai XY, O'Toole EA, Woodley DT: Ultraviolet A irradiation upregulates type VII collagen expression in human dermal fibroblasts. J Invest Dermatol 1997;108:125–128.

116 Soriani M, Rice-Evans C, Tyrrell RM: Modulation of the UVA activation of haem oxygenase, collagenase and cyclooxygenase gene expression by epigallocatechin in human skin cells. FEBS Lett 1998;439:253–257.

117 Huang RP, Fan Y, Boynton AL: UV irradiation upregulates Egr-1 expression at transcription level. J Cell Biochem 1999;73:227–236.

118 Park KC, Jung HC, Hwang JH, Youn SW, Ahn JS, Park SB, Kim KH, Chung JH, Youn JI: GM-CSF production by epithelial cell line: Upregulation by ultraviolet A. Photodermatol Photoimmunol Photomed 1997;13:133–138.

119 Gomer CJ, Luna M, Ferrario A, Rucker N: Increased transcription and translation of heme oxygenase in Chinese hamster fibroblasts following photodynamic stress or photofrin II incubation. Photochem Photobiol 1991;53:275–279.

120 Trautinger F, Kokesch C, Klosner G, Knobler RM, Kindas-Mugge I: Expression of the 72-kD heat shock protein is induced by ultraviolet A radiation in a human fibrosarcoma cell line. Exp Dermatol 1999;8:187–192.
121 Gollnick SO, Liu X, Owczarczak B, Musser DA, Henderson BW: Altered expression of interleukin 6 and interleukin 10 as a result of photodynamic therapy in vivo. Cancer Res 1997;57:3904–3909.
122 Morita A, Grewe M, Grether Beck S, Olaizola Horn S, Krutmann J: Induction of proinflammatory cytokines in human epidermoid carcinoma cells by in vitro ultraviolet A1 irradiation. Photochem Photobiol 1997;65:630–635.
123 Herrmann G, Wlaschek M, Lange TS, Prenzel K, Goerz G, Scharffetter-Kochanek K: UVA irradiation stimulates the synthesis of various matrix-metalloproteinases (MMPs) in cultured human fibroblasts. Exp Dermatol 1993;2:92–97.
123 Ahmad N, Feyes DK, Agarwal R, Mukhtar H: Photodynamic therapy results in induction of WAF1/CIP1/p21 leading to cell cycle arrest and apoptosis. Proc Natl Acad Sci USA 1998;95: 6977–6982.

Prof. Helmut Sies, Institut für Physiologische Chemie I, Heinrich-Heine-Universität Düsseldorf, Postfach 101007, D–40001 Düsseldorf (Germany)
Tel. +49 211 811 2707, Fax +49 211 811 3029, E-Mail Sies@uni-duesseldorf.de

Thiele J, Elsner P (eds): Oxidants and Antioxidants in Cutaneous Biology.
Curr Probl Dermatol. Basel, Karger, 2001, vol 29, pp 114–127

Reactive Oxygen Species as Mediators of UVB-Induced Mitogen-Activated Protein Kinase Activation in Keratinocytes

Dominik Peus[a], *Mark R. Pittelkow*[b]

[a] Department of Dermatology, Ludwig Maximilian University, Munich, Germany;
[b] Department of Dermatology, Mayo Clinic/Foundation, Rochester, Minn., USA

The increase in UVB, a minor but highly carcinogenic constituent of sunlight [1, 2], of the earth's surface has been linked to the recent stratospheric ozone depletion [3]. This has increased the interest of the effects of UVB on human skin, the natural target of ultraviolet radiation (UVR), since UVR is considered to be the primary cause for the vast majority of cutaneous malignancies [4]. UVR is the most efficient environmental carcinogen known and acts as both a tumor initiator as well as a tumor promoter [5–7]. Other chronic and acute effects of UVR on skin include erythema, inflammation, hyperpigmentation, hyperplasia and skin aging [8, 9]. These biological responses can be considered as consequences of a chain of signaling events that are triggered by the potent stimulus of UVR. This spectrum of electromagnetic radiation imparts photon energy that is translated into a mode of cellular communication with activation of intracellular signal transduction pathways. Previously, due to limited technical analysis, patterns of gene expression had been the major focus of investigation. However, it is now possible to elucidate earlier signaling events elicited by UVR in various cell types. The discoveries have been enabled by significant advances in our understanding of different cytoplasmic protein kinases and the introduction of specific antibodies and assays to analyze their state of 'activation' or phosphorylation.

Previous studies to investigate UV-induced signal transduction have used UVC irradiation, largely at supraphysiological doses as high as 2 orders of magnitude higher than those needed to induce gene transcription [10–13]. In contrast, the studies we initiated with human keratinocytes were designed to examine doses of 50–800 J/m^2 of UVB that are within the normal range of photon energies that produce a minimal erythema dose under standard skin testing conditions for skin types I–III and typically induce the physiological response of solar erythema [9]. Moreover, it should also be noted that effects of UVR on mitogen-activated protein kinase (MAPK) activation were largely examined in immortalized or carcinoma cell lines [11–14] where intracellular signal transduction pathways are frequently perturbed as part of the transformation process. Therefore, not only are the UVB doses we investigated relevant, but the serum-free cell culture system using normal human keratinocytes in early passage also provides additional physiological relevance in these findings to the in vivo state of correlating epidermis and skin [15, 16].

This review will focus on recent advances in mechanisms of UVB-induced signal transduction through MAPK cascades and particularly emphasizes findings obtained in vitro that most likely reflect the in vivo responses of epidermis and skin as closely as possible. Full elucidation of the pattern of activation of signaling cascades and their complex interactions, directly and indirectly, with nuclear targets that regulate gene expression increasingly emerges as a paradigm to predict cellular behavior. This may explain the intense interest in unraveling patterns of signaling pathway activation. However, the biological and chemical mediators of various cellular responses to UVR are only beginning to be identified and characterized.

Mitogen-Activated Protein Kinases Induced by UVB

One of the principal sets of gene products of the cell to transduce extracellular signals to the nucleus utilizes sequentially clustered protein kinase reactions. Approximately 1,200 genes have evolved in the mammalian cell that encode protein kinases [17]. Specific intracellular kinases are activated most frequently by receptor tyrosine kinase activation, which phosphorylate and activate other kinases, ultimately leading to activation of nuclear transcription factors that regulate gene transcription. MAPKs are a superfamily of enzymes that play a number of important roles in transmitting signals from the membrane or cytoplasm to the nucleus [18]. As cytoplasmic protein kinases, MAPKs mediate the transcription of many early genes and play a major role in modulating and coordinating these gene responses [19]. Three major MAPK signal transduction pathways have been identified and include the extracellular signal-regulated kinases (ERKs), c-Jun N-terminal kinases (JNKs) and p38 (also known as CSBP, RK, Mkp2) [20], as shown in figure 1. They are structurally related but biochemically and functionally distinct. ERKs, including the p42 and p44 MAPKs or ERK-1 and ERK-2, are the best studied and most distally situated enzymes in a three-kinase cascade. In this cascade,

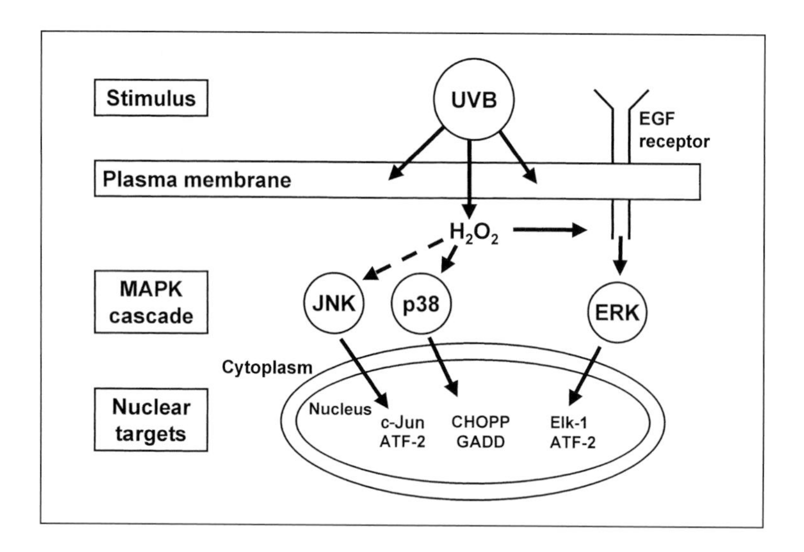

Fig. 1. Signaling pathway diagram: an overview of the scheme by which the three major MAPK pathways transmit signals from the membrane or cytosol to the nucleus. Whereas ERK is predominantly activated by growth factor ligand binding through growth factor receptors such as epidermal growth factor (EGF) receptor, activation of p38 and JNK are mainly stress induced. Sequential kinase activation leads to an orderly activation of these pathways which can be regarded as linear signaling cascades. Upon activation by different upstream protein kinases, these MAPKs translocate to the nucleus to activate specific transcription factors, e.g. ATF-2, CHOPP, Elk-1 or GADD. The role of H_2O_2 or other ROS in UV-induced and other signaling responses is only beginning to be elucidated in keratinocytes.

MAP/ERK kinase (MEK) kinase phosphorylates and activates MAPK/ERK kinases, MEK-1 and MEK-2, which in turn activate ERKs by dual phosphorylation on tyrosine and threonine residues within the protein kinase subdomain VIII [21]. Growth factors and phorbol esters primarily activate ERKs, whereas stress or inflammatory cytokines are only poor activators of ERKs [22, 23]. MAPK signaling responses have been shown to be central events for keratinocyte proliferation, clonal formation and survival [24]. Based on previous observations demonstrating that epidermal growth factor receptor (EGFR) activation is required for keratinocyte autocrine growth [25], we sought to identify mechanisms in UVB-induced ligand-independent EGFR and MAPK activation. Even though cellular stress has been considered to be a poor activator of ERKs we have recently shown that ERK-1/2 are rapidly and strongly activated within minutes of exposure to physiological doses of UVB in human keratinocytes [16].

Another member of the MAPK superfamily, p38, is also activated in a time-dependent and transient fashion by similar UVB doses. p38 has sequence homology to an enzyme in yeast termed HOG-1 (high-osmolarity glycerol response) [26] and is only poorly activated by epidermal growth factor and phorbol esters [27, 28]. The enzyme is activated by phosphorylation on tyrosine and threonine residues by a wide range of cellular stresses such as osmotic shock, heat shock, inflammatory cytokines and UV light [28–30]. Signals are transmitted via MEK kinases 3, 4 and 6 and small GTP-binding proteins such as Cdc42 and Rac [31]. p38 can be phosphorylated and activated by any of these MEK kinases [27, 32]. An expanding group of substrates for p38 have been identified including transcription factors such as ATF-2 [29] and CHOP-1 [33] and other protein kinases including MAPK-activated protein kinases 2 and 3 [28, 34]. Upon activation MAPK-activated protein kinase 2 can induce expression of the transcription factors CREB and ATF-1 [35].

The JNK or stress-activated protein kinase cascade participates in growth factor signaling in many cell types, but JNKs are mainly activated in response to various stress events that also activate p38 and include UVC, heat shock, osmotic imbalance, endotoxin and cytokines [36]. Like other members of the MAPK family, JNK/stress-activated protein kinase requires threonine and tyrosine phosphorylation for its enzymatic activity [29]. Upon activation, JNK translocates to the nucleus and induces transcription of AP-1-containing genes by specifically phosphorylating Ser 63 and Ser 73 of c-*jun* [37]. We have shown that physiological doses of UVB induce transient and rapid JNK activation within minutes of irradiation [38]. The distinct time courses and levels of activation and deactivation of MAPKs by UVB exposure indicate that these pathways are separate and regulated independently in keratinocytes.

Reactive Oxygen Species: Mediators of UVB-Induced Mitogen-Activated Protein Kinase Activation

Reactive oxygen species (ROS) are constitutively produced in cells by oxidative respiration [39]. Oxidative stress results when cellular levels of ROS exceed the counterregulatory antioxidant system and the redox balance within cells is perturbed. A dynamic flux in the intracellular redox state has been implicated as an important determinant in a host of cellular decision branch points ranging from proliferation to cell death [40]. Generation of ROS, including superoxide radical (O_2^-), hydrogen peroxide (O_2H_2) and hydroxyl radical (·OH), that exceeds a critical threshold has been associated with cutaneous aging as well as the pathogenesis of diseases such as cancer and inflammatory disorders [41–46]. UVR is a potent inducer of ROS·O_2^-, H_2O_2 and ·OH [47].

Based on previous assumptions of the relative potencies of UVA versus UVB in inducing cell stress, generation of ROS has been more extensively studied for UVA compared to UVB irradiation. In fibroblasts, production of ROS such as singlet oxygen and H_2O_2 had been reported in response to UVB [48, 49]. However, little was known about generation of ROS in keratinocytes following UVB, in part due to the lack of sufficiently sensitive assay techniques. Of interest, another short-lived metabolic product of oxygen, nitric oxide, was also found to be secreted in response to UVA or UVB in normal human keratinocytes [50]. Novel techniques have recently become available to precisely measure and quantitate the extracellular release of H_2O_2, such as Amplex, a derivative of dihydrophenoxazine [51]. Following UVB exposure in keratinocytes, we observed that extracellular levels of H_2O_2 were time-dependently induced, and these results were confirmed using the established fluorescent dye probe dihydrorhodamine, though dihydrorhodamine does not allow peroxide quantitation [15, 52]. Most recently, Schallreuter et al. [53] have shown in vivo, using Fourier transform Raman spectroscopy, that H_2O_2 is produced in the epidermis following UVB irradiation.

There is growing evidence that ROS serve as intracellular second messengers [40]. H_2O_2 is considered to be a physiological mediator of numerous cellular functions. It mimics insulin activity [54], regulates iron metabolism [55] and initiates apoptosis [56]. Recently, it has been shown that H_2O_2 induces ERK activation in various immortalized cell lines [57–59]. Several of our findings provide evidence that ROS, and in particular H_2O_2, function as prominent mediators in the activation and regulation of UVB-induced ERK-1/2 and p38 signal transduction pathways in normal human keratinocytes. Firstly, H_2O_2 itself is able to rapidly induce ERK and p38 activation, mimicking selected effects of UVB. Secondly, concentrations of exogenous H_2O_2 correlate with the intensity of MAPK phosphorylation and concentrations of intracellular H_2O_2 that are generated following UVB. Thirdly, the time course of UVB-induced generation of H_2O_2 correlates with the kinetics of ERK or p38 activation. Fourthly, specific antioxidants inhibit UVB- and H_2O_2-induced ERK and p38 activation [15–52].

However, neither the dose response correlation nor nonspecific antioxidants for H_2O_2 are definitive evidence to state that H_2O_2 is a second messenger required for UVB-induced EGFR phosphorylation. Clear distinction between the requirement for H_2O_2 in this process versus H_2O_2 as simply an associated cellular response, and the identification of H_2O_2 versus other ROS including hydroxyl radicals and superoxide anions in these cell responses, can only be established by decreasing endogenous levels of H_2O_2 in close proximity to the site of its production. A specific scavenger of H_2O_2 would offer a critical approach to solve this dilemma. To provide evidence for the requirement of

H_2O_2 in UVB-induced ERK-1/2 activation, we have shown that the H_2O_2-scavenging enzyme catalase can be overexpressed in keratinocytes by electroporation. Catalase-overexpressing keratinocytes strongly inhibit ERK-1/2 activation induced by UVB as well as by H_2O_2 and clearly demonstrate that H_2O_2 is an endogenous mediator required for EGFR-mediated ERK-1/2 activation induced by UVB or H_2O_2 [52].

Further investigation has demonstrated that the time course of ERK and p38 activation induced by H_2O_2 is similar in keratinocytes whereas clear differences are observed following UVB. This evidence suggests that factors in addition to H_2O_2 generation play a role in activation and, likely, dephosphorylation after UVB exposure, but these factors remain to be characterized. It can only be speculated whether other forms of oxidative stress such as lipid peroxidation also play a role in ERK or p38 activation following UVB. In this regard, lipid peroxidation of membrane lipids represents a degradative process, which is the consequence of the production and the propagation of free radical reactions primarily involving membrane polyunsaturated fatty acids. Interestingly, densitometric analysis to quantitate levels of UVB-induced JNK activation demonstrates that lipid peroxidation precedes but also correlates with JNK activation [15]. The specific roles of the complex chain reaction of lipid peroxidation for UV-induced signaling pathway activation remains to be further characterized.

Antioxidants in the Regulation of UVB-Induced Mitogen-Activated Protein Kinase Activation

In addition to characterizing the activation of ERK-1/2 and p38 pathways following UVB or H_2O_2 treatment, we examined potential mechanisms by which these pathways are activated and regulated by using different antioxidants. Cells contain an array of defense mechanisms to counteract and potentially prevent harmful overproduction of ROS. These extracellular protective mechanisms include enzymatic and chemical antioxidants which can also be specifically used to analyze mechanisms regulating the cellular stress response. Observations that the membrane-permeant antioxidant N-acetylcysteine inhibits UV-induced activation of EGFR [12, 60, 61] and ERKs [62] had already implicated a role for ROS in UV-induced signal transduction. Our experimental approach was based on the hypothesis that ROS are causally related mediators of UVB-induced MAPK activation, which implies that levels of activation should correlate in a dose response manner and generation of ROS precedes activation. ROS were measured following pretreatment of cells with structurally different antioxidants prior to UVB irradiation.

The antioxidants pyrrolidine dithiocarbamate or butyl hydroxyanisole inhibit ROS generation within minutes of UVB exposure in keratinocytes, and inhibition translates into attenuation of EGFR phosphorylation by these compounds [15]. The antioxidant ascorbic acid as vitamin C inhibits intracellular ROS as well as extracellular H_2O_2 production in response to UVB irradiation of human keratinocytes. Furthermore, levels of H_2O_2 in unstimulated human keratinocytes were also decreased in a concentration-dependent manner by ascorbic acid. These studies provide direct evidence that ascorbic acid decreases H_2O_2 levels in UVB-stimulated as well as nonstimulated keratinocytes. Reduction of ROS or H_2O_2 correlates with inhibition of EGFR phosphorylation and ERK-1/2 activation. Ascorbic acid also potently blocks ERK-1/2 and p38 activation by UVB and H_2O_2, whereas pyrrolidine dithiocarbamate and butyl hydroxyanisole are less effective. Pyrrolidine dithiocarbamate fails to inhibit UVB-induced p38 activation [15, 16]. These results indicate that the regulation of these signaling pathways by ROS is distinct, and cellular responses appear to differ depending on the types of ROS involved.

Role of UVB-Induced Extracellular Signal-Regulated Kinase Activation in Cell Survival

The functional role of the activation of different receptors and signal transduction pathways in response to UVR is only beginning to be unraveled. The role of ERK-1/2 activation in cell survival appears to depend on the cell type as well as the stimulus and the cellular environment. ERK-1/2 inhibition has no significant effect on the insulin- or brain-derived neurotrophic-factor-induced survival of cerebellar granule cells [63]. In cardiac myocytes, inhibition of ERKs increases apoptotic cell death [64] and cardiotrophin 1 promotes survival via activation of a signaling pathway requiring ERKs [65]. In hematopoietic cells, activation of ERK partially prevents apoptotic death [66]. Interestingly, increased resistance to H_2O_2 cytotoxicity was reported for NIH 3T3 cells expressing constitutively active MEK, the immediate upstream regulator of ERK. This study also showed that cells expressing kinase-defective MEK were more sensitive than cells expressing wild-type MEK [57].

The potent kinase inhibitor PD098059, a compound that specifically blocks activation of MEK-1 and MEK-2, both of which are upstream kinases of ERK, but not of p38 or JNK [67, 68], potently inhibited ERK-1/2 activation by UVB and H_2O_2 in human keratinocytes. No significant inhibition of p38 phosphorylation was observed [16]. Cell death increased after UVB exposure when ERK-1/2 activation was inhibited in keratinocytes by PD098059 [16]. These findings indicate that ERK-1/2 activation following UVB exposure plays a protective role against cellular stress and is a critical control point to determine keratinocyte survival as part of the UVB response. Whether activation of JNK and/or p38 is critical for promoting cell death in keratinocytes remains to be determined. Recent studies have shown that activation of JNK in immortalized cells appears to have been subverted to a growth-promoting, mitogenic pathway, where stress conditions favor initiated cells [69]. Based on these findings, it can be conjectured that the balance between pro- and antioxidative mechanisms within the cell as well as the transformed state will not only determine the activities of signal transduction pathways, but will also control the fate of the cell.

Mitogen-Activated Protein Kinase Activation: Dependence on Epidermal Growth Factor Receptor Phosphorylation

The mechanisms of activation of these MAPKs and their dependence on specific phosphorylation events of upstream kinases, such as receptor tyrosine kinases, particularly the EGFR, in mediating the stress response has been the subject of intense investigation. Mitogen-stimulated signal transduction leading to activation of ERK is initiated through the interaction of peptide growth factors with their receptor. Sachsenmaier et al. [13] were among the first to report that EGFR is phosphorylated by exposure to UVC. We have extended these findings to the serum-free keratinocyte culture system irradiated with physiological doses of UVB and observed that EGFR is also phosphorylated by UVB [70]. Though ligand-induced stimulation of receptor tyrosine kinases typically activates ERKs through the Ras/Raf-1/MEK cascade in many cell types [71, 72], the mechanism for UVB-induced and ligand-independent phosphorylation of EGFR in the early activation of cytoplasmic signaling pathways has been unclear.

UVR has been shown to activate Ras [13], and Ras has been reported to be necessary for ERK activation by H_2O_2 in various immortalized cell lines and smooth muscle cells [73]. More recently, suramin, which is known to inhibit ligand-receptor interactions and can inhibit ERK-2 activation by epidermal growth factor as well as by UVC irradiation [13], has been shown to block H_2O_2-induced ERK-2 activation [73]. In addition, overexpression of a dominant-negative mutant of Ras or Raf-1 abolishes the activation of transfected ERK-2 by H_2O_2 in cardiac myocytes, demonstrating that Ras and Raf-1 are critical for ERK activation by H_2O_2 [64]. To examine the dependence of ERK-1/2 activation on UVB-induced EGFR phosphorylation we have used the specific EGFR inhibitor PD153035 [70, 74] that potently blocks UVB-induced EGFR autophosphorylation and ERK-1/2 activation concentration dependently [75]. In addition, the neutralizing EGFR monoclonal antibody 528 effectively inhibits EGFR phosphorylation following UVB irradiation and also inhibits ERK-1/2 activation [52]. p38 activation was not affected by EGFR blockade [75]. These results demonstrate that ligand-independent EGFR autophosphorylation induced by UVB is required for ERK-1/2 activation. In contrast, UVB-induced p38 activation appears to be independent of EGFR phosphorylation. Inhibition of UVB-induced EGFR phosphorylation by PD153035 also failed to block JNK activation [76]. Thus, ERK-1/2 but not p38 or JNK pathway activation is dependent on EGFR phosphorylation following UVB irradiation in cultured keratinocytes.

Phosphatases: Regulators of Mitogen-Activated Protein Kinase

The kinetics of time-dependent, UVB-induced activation of ERK-1/2 and p38 are distinct with early and rapid downregulation of ERK-1/2 activation within minutes of stimulation compared to more sustained, prolonged activation of p38 [16]. What are the mechanisms of downregulation that result in these pronounced differences in the phosphorylation status of these related MAPKs? Upon normal stimulation, MAPKs translocate into the nucleus where they selectively activate several different transcription factors and are then rapidly inactivated by dual specific phosphatases [33, 77–79]. Dephosphorylation of activated ERK-1/2 and p38 is mediated by the recently isolated and characterized family of dual-specificity (Thr/Tyr) MAPK phosphatases (MKPs). Though the discovery of the plethora of MKPs has lagged behind that of MAPKs, it is clear that MKPs perform crucial functions in orchestrating the cellular stress response. Up to nine distinct mammalian MKPs have been identified to date [80]. MKP-1 to MKP-4 [81–83] are known to be induced by cellular stress and rapidly localize to the nucleus [84]. MKP-1, -2 and -4 appear to predominantly inactivate JNKs and p38 MAPKs [83, 85] whereas MKP-3 has unique specificity for ERK [80].

The body of evidence draws together MKPs as a model for the inactivation of ERKs and other MAPKs. Specific sites of inactivation have recently been suggested as a mechanism to further target substrate-enzyme interactions [80]. MKPs, such as MKP-1/2 or PAC-1, dephosphorylate ERK in the nucleus. In contrast, MKP-3 can bind to ERK and inactivate ERK in the cytosol. Under certain conditions, other types of phosphatases, such as PP2A protein phosphatase or unidentified PTPase protein tyrosine phosphatase, are also responsible for the inactivation of ERK [80, 86]. The rapid, yet transient activation of ERK by UVB radiation highlights the dynamic nature of intracellular phosphorylation levels and underscores a role for phosphatases in finely regulating this response. Further investigation will be required to more precisely delineate the mechanisms and identify the host of specific regulators in stress-induced MAPK inactivation.

Perspective

Evidence is accumulating from our laboratories and others that the balance between pro- and antioxidative status within the keratinocyte determines to a significant extent the levels of activation of downstream signaling cascades, such as EGFR/ERK-1/2 and p38 that are differentially triggered in response to UVB radiation. ROS, in particular H_2O_2, as well as phosphatases play critical roles as regulators of ERK-1/2 and p38 pathways and mediate various cellular responses following UVB exposure. ERK-1/2 primarily functions to protect keratinocytes from cell death after UVB radiation and, likely, other forms of oxidative stress. Early activation of diverse signaling pathways in response to UVR may also be involved in erythema and the inflammatory reaction as well as photoaging, photodermatoses and carcinogenesis. Therefore, a more complete understanding of the activation cascades in the network of signaling pathways and the generation of ROS within keratinocytes will be needed to develop effective chemopreventive agents, protective topical formulations and possibly systemic agents, to counteract the deleterious effects of UVR.

Acknowledgment

This work was supported by the Mayo Foundation and the German Research Foundation (DFG, PE 635/2-1). D.P. was supported by the Theodor-Nasemann-Stipendium. We thank Lilette Peus for secretarial assistance.

References

1 Brash DE, Rudolph JA, Simon JA, Lin A, McKenna GJ, Baden HP, Halperin AJ, Ponten J: A role for sunlight in skin cancer: UV-induced p53 mutations in squamous cell carcinoma. Proc Natl Acad Sci USA 1991;88:10124–10128.

2 Tornaletti S, Pfeifer GP: UV damage and repair mechanisms in mammalian cells. Bioessays 1996; 18:221–228.

3 Madronich S, de Gruijl FR: Skin cancer and UV radiation. Nature 1993;366:23.

4 Miller DL, Weinstock MA: Nonmelanoma skin cancer in the United States: Incidence. J Am Acad Dermatol 1994;30:774–778.

5 Epstein JH: Photocarcinogenesis: A review. Natl Cancer Inst Monogr 1978:13–25.

6 Romerdahl CA, Stephens LC, Bucana C, Kripke ML: The role of ultraviolet radiation in the induction of melanocytic skin tumors in inbred mice. Cancer Commun 1989;1:209–216.

7 Matsui MS, DeLeo VA: Longwave ultraviolet radiation and promotion of skin cancer. Cancer Cells 1991;3:8–12.

8 Fisher GJ, Wang ZQ, Datta SC, Varani J, Kang S, Voorhees JJ: Pathophysiology of premature skin aging induced by ultraviolet light. N Engl J Med 1997;337:1419–1428.

9 Pathak MA, Nghiem P, Fitzpatrick TB: Acute and chronic effects of the sun; in Freedberg IM, Eisen AZ, Wolff K, Austen KF, Goldsmith LA, Katz SI, Fitzpatrick TB (eds): Fitzpatrick's Dermatology in General Medicine. New York, McGraw-Hill, 1999.

10 Bender K, Blattner C, Knebel A, Iordanov M, Herrlich P, Rahmsdorf HJ: UV-induced signal transduction. J Photochem Photobiol B 1997;37:1–17.

11 Iordanov M, Bender K, Ade T, Schmid W, Sachsenmaier C, Engel K, Gaestel M, Rahmsdorf HJ, Herrlich P: CREB is activated by UVC through a p38/HOG-1-dependent protein kinase. EMBO J 1997;16:1009–1022.

12 Huang RP, Wu JX, Fan Y, Adamson ED: UV activates growth factor receptors via reactive oxygen intermediates. J Cell Biol 1996;133:211–220.

13 Sachsenmaier C, Radler-Pohl A, Zinck R, Nordheim A, Herrlich P, Rahmsdorf HJ: Involvement of growth factor receptors in the mammalian UVC response. Cell 1994;78:963–972.

14 Rosette C, Karin M: Ultraviolet light and osmotic stress: Activation of the JNK cascade through multiple growth factor and cytokine receptors. Science 1996;274:1194–1197.

15 Peus D, Vasa A, Meves A, Pott M, Beyerle A, Squillace K, Pittelkow MR: H_2O_2 is an important mediator of UVB-induced EGF-receptor phosphorylation in cultured keratinocytes. J Invest Dermatol 1998;110:966–971.

16 Peus D, Vasa RA, Beyerle A, Meves A, Krautmacher C, Pittelkow MR: UVB activates ERK1/2 and p38 signaling pathways via reactive oxygen species in cultured keratinocytes. J Invest Dermatol 1999;112:751–756.

17 Hunter T: 1,001 protein kinases redux – Towards 2000. Semin Cell Biol 1994;5:367–376.

18 Seger R, Krebs EG: The MAPK signaling cascade. Faseb J 1995;9:726–735.

19 Karin M: The regulation of AP-1 activity by mitogen-activated protein kinases. J Biol Chem 1995; 270:16483–16486.

20 Su B, Karin M: Mitogen-activated protein kinase cascades and regulation of gene expression. Curr Opin Immunol 1996;8:402–411.

21 Marshall CJ: MAP kinase kinase kinase, MAP kinase kinase and MAP kinase. Curr Opin Genet Dev 1994;4:82–89.

22 Whitmarsh AJ, Shore P, Sharrocks AD, Davis RJ: Integration of MAP kinase signal transduction pathways at the serum response element. Science 1995;269:403–407.

23 Xia Z, Dickens M, Raingeaud J, Davis RJ, Greenberg ME: Opposing effects of ERK and JNK-p38 MAP kinases on apoptosis. Science 1995;270:1326–1331.

24 Geilen CC, Wieprecht M, Orfanos CE: The mitogen-activated protein kinases system (MAP kinase cascade): Its role in skin signal transduction. A review. J Dermatol Sci 1996;12:255–262.

25 Pittelkow MR, Cook PW, Shipley GD, Derynck R, Coffey RJJ: Autonomous growth of human keratinocytes requires epidermal growth factor receptor occupancy. Cell Growth Differ 1993;4:513–521.

26 Brewster JL, de Valoir T, Dwyer ND, Winter E, Gustin MC: An osmosensing signal transduction pathway in yeast. Science 1993;259:1760–1763.

27 Derijard B, Raingeaud J, Barrett T, Wu IH, Han J, Ulevitch RJ, Davis RJ: Independent human MAP-kinase signal transduction pathways defined by MEK and MKK isoforms. Science 1995;267: 682–685.

28 Rouse J, Cohen P, Trigon S, Morange M, Alonso-Llamazares A, Zamanillo D, Hunt T, Nebreda AR: A novel kinase cascade triggered by stress and heat shock that stimulates MAPKAP kinase-2 and phosphorylation of the small heat shock proteins. Cell 1994;78:1027–1037.

29 Raingeaud J, Gupta S, Rogers JS, Dickens M, Han J, Ulevitch RJ, Davis RJ: Pro-inflammatory cytokines and environmental stress cause p38 mitogen-activated protein kinase activation by dual phosphorylation on tyrosine and threonine. J Biol Chem 1995;270:7420–7426.

30 Wesselborg S, Bauer MKA, Vogt M, Schmitz ML, Schulze-Osthoff K: Activation of transcription factor NF-kappaB and p38 mitogen-activated protein kinase is mediated by distinct and separate stress effector pathways. J Biol Chem 1997;272:12422–12429.

31 Lamarche N, Tapon N, Stowers L, Burbelo PD, Aspenstrom P, Bridges T, Chant J, Hall A: Rac and Cdc42 induce actin polymerization and G1 cell cycle progression independently of p65PAK and the JNK/SAPK MAP kinase cascade. Cell 1996;87:519–529.
32 Raingeaud J, Whitmarsh AJ, Barrett T, Derijard B, Davis RJ: MKK3- and MKK6-regulated gene expression is mediated by the p38 mitogen-activated protein kinase signal transduction pathway. Mol Cell Biol 1996;16:1247–1255.
33 Wang XZ, Ron D: Stress-induced phosphorylation and activation of the transcription factor Chop (Gadd153) by p38 MAP kinase. Science 1996;272:1347–1349.
34 McLaughlin MM, Kumar S, McDonnell PC, Van Horn S, Lee JC, Livi GP, Young PR: Identification of mitogen-activated protein (MAP) kinase-activated protein kinase-3, a novel substrate of CSBP p38 MAP kinase. J Biol Chem 1996;271:8488–8492.
35 Tan Y, Rouse J, Zhang A, Cariati S, Cohen P, Comb MJ: FGF and stress regulate CREB and ATF-1 via a pathway involving p38 MAP kinase and MAPKAP kinase-2. EMBO J 1996;15: 4629–4642.
36 Canman CE, Kastan MB: Signal transduction: Three paths to stress relief. Nature 1996;384:213–214.
37 Cui XL, Douglas JG: Arachidonic acid activates c-*jun* N-terminal kinase through NADPH oxidase in rabbit proximal tubular epithelial cells. Proc Natl Acad Sci USA 1997;94:3771–3776.
38 Peus D, Vasa A, Beyerle A, Meves A, Pittelkow MR: UVB-induced JNK activation is mediated by lipid peroxidation in keratinocytes. J Invest Dermatol 1998;110:691.
39 Cerutti PA: Prooxidant states and tumor promotion. Science 1985;227:375–381.
40 Finkel T: Signal transduction by reactive oxygen species in non-phagocytic cells. J Leukoc Biol 1999;65:337–340.
41 Black HS: Potential involvement of free radical reactions in ultraviolet light-mediated cutaneous damage. Photochem Photobiol 1987;46:213–221.
42 Halliwell B: Free radicals, reactive oxygen species and human disease: A critical evaluation with special reference to atherosclerosis. Br J Exp Pathol 1989;70:737–757.
43 Fridovich I: The biology of oxygen radicals. Science 1978;201:875–880.
44 Floyd RA: Role of oxygen free radicals in carcinogenesis and brain ischemia. FASEB J 1990;4: 2587–2597.
45 Janssen YM, Van Houten B, Borm PJ, Mossman BT: Cell and tissue responses to oxidative damage. Lab Invest 1993;69:261–274.
46 Tyrrell RM: The molecular and cellular pathology of solar ultraviolet radiation. Mol Aspects Med 1994;15:1–77.
47 Darr D, Fridovich I: Free radicals in cutaneous biology. J Invest Dermatol 1994;102:671–675.
48 Tyrrell RM, Pidoux M: Singlet oxygen involvement in the inactivation of cultured human fibroblasts by UVA (334 nm, 365 nm) and near-visible (405 nm) radiations. Photochem Photobiol 1989;49: 407–412.
49 Vile GF, Tyrrell RM: UVA radiation-induced oxidative damage to lipids and proteins in vitro and in human skin fibroblasts is dependent on iron and singlet oxygen. Free Radic Biol Med 1995;18:721–730.
50 Romero-Graillet C, Aberdam E, Clement M, Ortonne JP, Ballotti R: Nitric oxide produced by ultraviolet-irradiated keratinocytes stimulates melanogenesis. J Clin Invest 1997;99:635–642.
51 Mohanty JG, Jaffe JS, Schulman ES, Raible DG: A highly sensitive fluorescent micro-assay of H_2O_2 release from activated human leukocytes using a dihydroxyphenoxazine derivative. J Immunol Methods 1997;202:133–141.
52 Peus D, Meves A, Vasa R, Beyerle A, O'Brien T, Pittelkow MR: H_2O_2 is required for UVB-induced EGF receptor and downstream signaling pathway activation. Free Radic Biol Med 1999; 27:1197–1202.
53 Schallreuter KU, Moore J, Wood JM, Beazley WD, Gaze DC, Tobin DJ, Marshall HS, Panske A, Panzig E, Hibberts NA: In vivo and in vitro evidence for hydrogen peroxide (H_2O_2) accumulation in the epidermis of patients with vitiligo and its successful removal by a UVB-activated pseudocatalase. J Investig Dermatol Symp Proc 1999;4:91–96.
54 Mukherjee SP, Mukherjee C, Lynn WS: Activation of pyruvate dehydrogenase in rat adipocytes by concanavalin A: Evidence for insulin-like effect mediated by hydrogen peroxide. Biochem Biophys Res Commun 1980;93:36–41.

55 Pantopoulos K, Hentze MW: Rapid responses to oxidative stress mediated by iron regulatory protein. EMBO J 1995;14:2917–2924.

56 Hockenbery DM, Oltvai ZN, Yin XM, Milliman CL, Korsmeyer SJ: Bcl-2 functions in an antioxidant pathway to prevent apoptosis. Cell 1993;75:241–251.

57 Guyton KZ, Liu Y, Gorospe M, Xu Q, Holbrook NJ: Activation of mitogen-activated protein kinase by H_2O_2: Role in cell survival following oxidant injury. J Biol Chem 1996;271:4138–4142.

58 Rao GN, Glasgow WC, Eling TE, Runge MS: Role of hydroperoxyeicosatetraenoic acids in oxidative stress-induced activating protein 1 (AP-1) activity. J Biol Chem 1996;271:27760–27764.

59 Zhao Z, Tan Z, Diltz CD, You M, Fischer EH: Activation of mitogen-activated protein (MAP) kinase pathway by pervanadate, a potent inhibitor of tyrosine phosphatases. J Biol Chem 1996; 271:22251–22255.

60 Devary Y, Gottlieb RA, Smeal T, Karin M: The mammalian ultraviolet response is triggered by activation of Src tyrosine kinases. Cell 1992;71:1081–1091.

61 Knebel A, Rahmsdorf HJ, Ullrich A, Herrlich P: Dephosphorylation of receptor tyrosine kinases as target of regulation by radiation, oxidants or alkylating agents. EMBO J 1996;15:5314–5325.

62 Assefa Z, Garmyn M, Bouillon R, Merlevede W, Vandenheede JR, Agostinis P: Differential stimulation of ERK and JNK activities by ultraviolet B irradiation and epidermal growth factor in human keratinocytes. J Invest Dermatol 1997;108:886–891.

63 Gunn-Moore FJ, Williams AG, Toms NJ, Tavare JM: Activation of mitogen-activated protein kinase and p70S6 kinase is not correlated with cerebellar granule cell survival. Biochem J 1997;324:365–369.

64 Aikawa R, Komuro I, Yamazaki T, Zou Y, Kudoh S, Tanaka M, Shiojima I, Hiroi Y, Yazaki Y: Oxidative stress activates extracellular signal-regulated kinases through Src and Ras in cultured cardiac myocytes of neonatal rats. J Clin Invest 1997;100:1813–1821.

65 Sheng Z, Knowlton K, Chen J, Hoshijima M, Brown JH, Chien KR: Cardiotrophin 1 (CT-1) inhibition of cardiac myocyte apoptosis via a mitogen-activated protein kinase-dependent pathway: Divergence from downstream CT-1 signals for myocardial cell hypertrophy. J Biol Chem 1997;272: 5783–5791.

66 Kinoshita T, Shirouzu M, Kamiya A, Hashimoto K, Yokoyama S, Miyajima A: Raf/MAPK and rapamycin-sensitive pathways mediate the anti-apoptotic function of p21Ras in IL-3-dependent hematopoietic cells. Oncogene 1997;15:619–627.

67 Alessi DR, Cuenda A, Cohen P, Dudley DT, Saltiel AR: PD 098059 is a specific inhibitor of the activation of mitogen-activated protein kinase kinase in vitro and in vivo. J Biol Chem 1995;270: 27489–27494.

68 Dudley DT, Pang L, Decker SJ, Bridges AJ, Saltiel AR: A synthetic inhibitor of the mitogen-activated protein kinase cascade. Proc Natl Acad Sci USA 1995;92:7686–7689.

69 Bost F, McKay R, Dean N, Mercola D: The JUN kinase/stress-activated protein kinase pathway is required for epidermal growth factor stimulation of growth of human A549 lung carcinoma cells. J Biol Chem 1997;272:33422–33429.

70 Peus D, Hamacher L, Pittelkow MR: EGF-receptor tyrosine kinase inhibition induces keratinocyte growth arrest and terminal differentiation. J Invest Dermatol 1997;109:751–756.

71 Rozakis-Adcock M, McGlade J, Mbamalu G, Pelicci G, Daly R, Li W, Batzer A, Thomas S, Brugge J, Pelicci PG, et al: Association of the Shc and Grb2/Sem5 SH2-containing proteins is implicated in activation of the Ras pathway by tyrosine kinases. Nature 1992;360:689–692.

72 Sadoshima J, Izumo S: The heterotrimeric G q protein-coupled angiotensin II receptor activates p21 *ras* via the tyrosine kinase-Shc-Grb2-Sos pathway in cardiac myocytes. EMBO J 1996;15: 775–787.

73 Guyton KZ, Yusen L, Myriam G, Qingo X, Nikki JH: Activation of mitogen-activated protein kinase H_2O_2. J Biol Chem 1995;271:4138–4142.

74 Fry DW, Kraker AJ, McMichael A, Ambroso LA, Nelson JM, Leopold WR, Connors RW, Bridges AJ: A specific inhibitor of the epidermal growth factor receptor tyrosine kinase. Science 1994;265: 1093–1095.

75 Vasa R, Peus D, Beyerle A, Pittelkow MR: Reactive oxygen species are important mediators of UVB-induced ERK1/2 and p38 activation in cultured keratinocytes. J Invest Dermatol 1998;110: 690.

76 Vasa R, Peus D, Kalli K, Pittelkow MR: UVB and H_2O_2 induce JNK activation and EGF receptor phosphorylation: Potential roles in keratinocyte survival. J Invest Dermatol 1997;108:619.

77 Keyse SM, Emslie EA: Oxidative stress and heat shock induce a human gene encoding a protein-tyrosine phosphatase. Nature 1992;359:644–647.

78 Ward Y, Gupta S, Jensen P, Wartmann M, Davis RJ, Kelly K: Control of MAP kinase activation by the mitogen-induced threonine/tyrosine phosphatase PAC1. Nature 1994;367:651–654.

79 Sun H, Charles CH, Lau LF, Tonks NK: MKP-1 (3CH134), an immediate early gene product, is a dual specificity phosphatase that dephosphorylates MAP kinase in vivo. Cell 1993;75:487–493.

80 Haneda M, Sugimoto T, Kikkawa R: Mitogen-activated protein kinase phosphatase: A negative regulator of the mitogen-activated protein kinase cascade. Eur J Pharmacol 1999;365:1–7.

81 Chu Y, Solski PA, Khosravi-Far R, Der CJ, Kelly K: The mitogen-activated protein kinase phosphatases PAC1, MKP-1, and MKP-2 have unique substrate specificities and reduced activity in vivo toward the ERK2 sevenmaker mutation. J Biol Chem 1996;271:6497–6591.

82 Muda M, Boschert U, Dickinson R, Martinou JC, Camps M, Schlegel W, Arkinstall S: Mkp-3, a novel cytosolic protein-tyrosine phosphatase that exemplifies a new class of mitogen-activated protein kinase phosphatase. J Biol Chem 1996;271:4319–4326.

83 Muda M, Boschert U, Smith A, Antonsson B, Gillieron C, Chabert C, Camps M, Martinou I, Ashworth A, Arkinstall S: Molecular cloning and functional characterization of a novel mitogen-activated protein kinase phosphatase, Mkp-4. J Biol Chem 1997;272:5141–5151.

84 Keyse SM: An emerging family of dual specificity MAP kinase phosphatases. Biochim Biophys Acta 1995;1265:152–160.

85 Hirsch DD, Stork PJS: Mitogen-activated protein kinase phosphatases inactivate stress-activated protein kinase pathways in vivo. J Biol Chem 1997;272:4568–4575.

86 Alessi DR, Gomez N, Moorhead G, Lewis T, Keyse SM, Cohen P: Inactivation of p42 MAP kinase by protein phosphatase 2A and a protein tyrosine phosphatase, but not CL100, in various cell lines. Curr Biol 1995;5:283–295.

Dominik Peus, MD, Department of Dermatology, Ludwig-Maximilians-Universität München, Frauenlobstrasse 9–11, D–80337 München (Germany)
Tel. +49 89 46204695, Fax +49 89 46202503, E-Mail D.Peus@lrz.uni-muenchen.de

Thiele J, Elsner P (eds): Oxidants and Antioxidants in Cutaneous Biology.
Curr Probl Dermatol. Basel, Karger, 2001, vol 29, pp 128–139

Antioxidants in Chemoprevention of Skin Cancer

Nihal Ahmad, Santosh K. Katiyar, Hasan Mukhtar

Department of Dermatology, Case Western Reserve University, Cleveland, Ohio, USA

Alarmingly increasing incidences of skin cancer are being reported from many countries where the majority of the population is Caucasian. According to an estimate by the American Cancer Society, approximately 1.3 million cases of basal-cell or squamous-cell cancers are diagnosed annually in the USA alone [1]. Some more disturbing facts about skin cancer are: (1) it attacks 1 out of every 7 Americans each year making it the most prevalent form of cancer [1], and (2) recent epidemiological studies have shown that there is an increased risk of other lethal cancer types in the individuals with a history of skin cancer [2]. It is also important to mention here that among all the cancers, skin cancer is believed to be one of the most preventable and/or curable cancer types [1]. The most serious form of skin cancer is melanoma, which is expected to be diagnosed in about 47,700 persons in the year 2000 [1]. Since the early 1970s, the incidence rate of melanoma has increased significantly, on average 4% per year, from 5.7/100,000 in 1973 to 13.8/100,000 in 1996 [1]. Other important skin cancers include Kaposi's sarcoma and cutaneous T-cell lymphoma.

The reactive oxygen intermediates (ROI) such as superoxide anion, hydrogen peroxide and singlet oxygen are believed to play a major role in many pathological conditions including skin cancer [3]. The ROI are very short-lived species and can react with DNA protein and unsaturated fatty acids, thereby causing DNA strand breaks and oxidative damage as well as protein-protein DNA cross-links [3]. The oxidation of lipids produces lipid peroxides, which are comparatively longer-lived species and can initiate the chain reactions to enhance the oxidative damage. In the body, the ROI and the lipid peroxides are generally produced by xenobiotic metabolism and through the respiratory chain [3]. In the skin, in addition to these normal sources, the ROI are also produced by the UV radiations from sun exposure [3]. The damages caused by these ROI may result in many skin disorders including skin cancer.

The skin is known to possess a variety of antioxidant mechanisms to defend it from damages caused by the ROI [3]. However, an excessive exposure to the ROI may render the defense system incapable of coping with the damaging effects, thereby leading to pathological conditions including skin cancer. In this scenario, exogenous supplementation of the antioxidants may provide protection against skin cancer. This approach is known as chemoprevention, which by definition is a means of cancer control in which the occurrence of the disease can be entirely prevented, slowed or reversed by the administration of one or more naturally occurring and/or synthetic compounds [4]. The expanded definition of cancer chemoprevention also includes the chemotherapy of precancerous lesions. Chemoprevention differs from cancer treatment in that the goal of this approach is to lower the rate of cancer incidence. In recent years, the naturally occurring compounds, especially the antioxidants, present in the common diet and beverages consumed by human populations have gained considerable attention as chemopreventive agents for a potential human benefit [4]. Supplementation of such agents in skin care products could lead to chemopreventive effects in the skin. This chapter will discuss the use of antioxidants for the prevention and possibly the treatment of skin cancer. We will specifically focus our discussion on naturally occurring antioxidants present in the diet and beverages.

Skin Cancer Development and Antioxidants

The development of skin cancer is a complex multistage process that is best explained by a three-step initiation-promotion-progression system that is mediated via various cellular, tissue, biochemical and molecular changes [3–8 and references therein]. The mouse skin model of multistage carcinogenesis has provided a conceptual framework for epithelial carcinogenesis mechanisms for many years. Initiation is the first step in multistage skin carcinogenesis that involves carcinogen-induced genetic changes [5]. The second step is the promotion stage, which involves many processes where the initiated cells undergo selective clonal expansion to form visible premalignant lesions known as papillomas [3–8 and references therein]. The progression stage involves the conversion of papillomas to malignant tumors. A detailed account of this process has been reviewed earlier [3–8 and references therein].

The carcinogens and tumor promoters, directly or indirectly, may generate ROI that are counteracted by the endogenous antioxidants in an effective manner [3–8 and references therein]. Accumulating evidence has suggested that ROI are important in all stages of skin cancer development [3–8 and references therein]. In the body, under normal physiological conditions, there is a balance between these prooxidant and antioxidant species. Under certain situations, such as the exposure to carcinogens, this balance shifts towards prooxidant, i.e. there is much increased generation of ROI, which could not be counteracted by the available antioxidants in the system. This situation may lead to the development of skin cancer. Many studies have shown that chemoprevention by exogenous antioxidants may prevent against skin cancer [6 and references therein].

Skin Cancer Chemoprevention by Naturally Occurring Antioxidants

In recent years, the naturally occurring compounds, especially the antioxidants, present in the common diet and beverages consumed by human populations, have gained considerable attention as chemopreventive agents against many cancers including skin cancer [6 and references therein]. It is being increasingly appreciated that chemoprevention may be a plausible strategy for the management of skin cancer [4, 6 and references therein]. Studies from this laboratory and others worldwide have shown that naturally occurring compounds present in human diet/beverages such as green tea polyphenols, resveratrol, curcumin, silymarin, ginger and diallyl sulfide afford protection against the development of skin cancer, both under in vitro (in culture system) as well as under in vivo (in animal models) situations. Furthermore, this approach appears to have practical implications in reducing the skin cancer risk because unlike the carcinogenic environmental factors that are difficult to control, individuals can modify their dietary habits, use of skin care products and lifestyle.

Skin Cancer Chemoprevention by Green Tea Polyphenols

Studies from our laboratory and from many other laboratories around the world have shown the chemopreventive potential of antioxidant polyphenols present in green tea against skin cancer [7 and references therein]. Tea is obtained from the leaves of the plant Camellia sinensis and is a popularly consumed beverage in the world [7 and references therein]. Tea is available in many forms, and the widely used forms of tea include black tea (78%, mainly consumed in Western countries and in some Asian countries), green tea (20%, consumed mainly by the Asian population) and oolong tea (2%, limited to Southeastern China) [7]. Most of the skin cancer chemoprevention studies on tea have been done with green tea while few studies have also investigated the potential of black tea against skin cancer. Green tea contains many polyphenolic antioxidants, which are thought to be responsible for its chemopreventive potential [7 and references therein]. The major polyphenolic antioxidants present in green tea are (–)-epicatechin, (–)-epigallocatechin, (–)-epicatechin-3-gallate and (–)-epigallocatechin-3-gallate (EGCG). Based on many studies, EGCG is believed to be the most potential antioxidant in green tea [7 and references therein]. In fact, studies have shown that the antioxidant activity of EGCG is much higher than that of the well-known antioxidants vitamin E and vitamin C [8].

In many mouse skin models, the topical application or oral consumption of a polyphenolic mixture obtained from green tea, referred to as green tea polyphenols (GTP), has been shown to afford protection against chemical- as well as UVB-induced skin carcinogenesis and inflammatory responses [7, 9, 10].

Studies have shown that GTP affords protection against erythema, edema and hyperplastic epithelial responses, which are often regarded as early markers of skin tumor formation [9, 10 and references therein]. Topical application of GTP to mouse skin has been shown to protect against 12-O-tetradecanoylphorbol-13-acetate (TPA)-induced epidermal edema, erythema, hyperplasia, infiltration of leukocytes and induction of cyclooxygenase and lipoxygenase activities in mouse skin [9, 10 and references therein]. Topical application of GTP also inhibited the induction of tumor promoters TPA-, mezerein- and benzoyl-peroxide-mediated upregulation of interleukin (IL)-1α mRNA [reviewed in 10].

UV-induced skin injury and oxidative stress have been associated with many dermatological disorders including skin cancer [3]. Studies have shown that oral feeding of GTP to Skh-1 hairless mice and a sole source of drinking fluid resulted in significant protection against UV-induced edema, erythema and depletion of antioxidant defense enzyme systems [reviewed in 10]. Topical application of GTP on mouse skin prior to UV exposure (72 mJ/cm^2) also resulted in a decreased UV-induced hyperplastic response, myeloperoxidase activity, infiltration of inflammatory leukocytes and inhibition of the contact hypersensitivity response [reviewed in 10].

In a recent study from this laboratory, the relevance of the extensive in vitro and in vivo laboratory data showing the preventive effects of GTP against UV-radiation-mediated damages to humans was sought [9]. In this study, Katiyar et al. [9] demonstrated that a topical application of EGCG (3 $mg/2.5\ cm^2$) prior to the 4 MED of UVB radiation resulted in significant inhibition of UVB-caused infiltration of leukocytes, myeloperoxidase activity and erythema in human skin. These infiltrating leukocytes are regarded as the major source of ROI. Further, its was also shown that skin treatment of EGCG prior to UVB resulted in a significant reduction of UVB-mediated prostaglandin metabolites, i.e. prostaglandin E_2 formation in the microsomes [9]. The prostaglandin metabolites are known to play a key role in the generation of ROI and skin tumor promotion in multistage skin carcinogenesis. Based on this study, it was suggested that EGCG may be a useful antioxidant for the protection against UVB responses including photocarcinogenesis [9]. These observations provided possible mechanisms involved in the anti-inflammatory effects of green tea.

Many studies have shown that the oral feeding of GTP resulted in significant protection against skin tumorigenesis in many animal models such as SENCAR, CD-1 and Balb/C mice [reviewed in 7, 10]. Topical application of GTP to Balb/C mice for 1 week, prior to the application of 3-methylcholanthrene, was found to result in significant protection against the development of skin tumors [reviewed in 7, 10]. Similarly, topical application of GTP has been shown to result in significant protection against 7,12-dimethylbenz[a]-anthracene (DMBA)-initiated TPA-promoted skin tumorigenesis both in terms of tumor incidence and the multiplicity of tumors [reviewed in 7, 10]. Studies have shown that the topical application of GTP or EGCG inhibited tumor promotion by TPA, teleocidin and okadaic acid [reviewed in 7, 10].

Several studies have shown that oral feeding of GTP via the drinking water results in significant protection against UVB-mediated skin tumorigenesis in the Skh-1 hairless mouse [reviewed in 9, 10]. In another study, Gensler et al. [reviewed in 10] demonstrated that topically applied EGCG resulted in significant inhibition of photocarcinogenesis in Balb/CAnNHsd mice. The water extract of green tea, given as a sole source of drinking fluid to mice, was found to afford prevention of UVB-radiation-mediated tumor initiation and tumor promotion in female CD-1 mice [reviewed in 10]. In this study, water extract of green tea was also shown to impart partial regression of established skin papillomas in mice.

Studies have suggested that UVB-radiation-mediated modulation of cytokines may play a key role in UVB responses including photocarcinogenesis [9, 10 and references therein]. Studies have also shown that a variety of skin cancers produce high levels of IL-10 and that the UV-mediated induction of IL-10 may be blocked by neutralizing antibodies against IL-10 or by antioxidants, which may impart protection against immunosuppression and skin tumor formation [9, 10 and references therein]. Studies from this laboratory have shown that topical application of EGCG prior to a single dose of UV (72 mJ/cm^2) exposure to Skh-1 hairless mice resulted in (1) reversal of UV-induced immunosuppression to a contact sensitizer, (2) reduced UV-induced production of IL-10 in skin as well as in draining lymph nodes and (3) significantly increased production of IL-12 in draining lymph nodes [10 and references therein]. This study suggested the possible application of green tea or its polyphenols to protect against UV-radiation-induced adverse effects on the immune system.

Skin Cancer Chemoprevention by Resveratrol

Resveratrol, chemically known as trans-3,5,4′-trihydroxystilbene, is a naturally occurring polyphenolic antioxidant present in grapes, berries, peanuts and red wine [11 and references therein]. In plants, resveratrol acts as a phytoalexin that protects against fungal infections. The cancer chemopreventive properties of resveratrol were first demonstrated when, in a mouse model of chemical carcinogenesis, this antioxidant was found to possess skin cancer chemopreventive activity [11 and references therein]. In this study, resveratrol was found to be effective against all the three major stages of carcinogenesis, i.e. initiation, promotion and progression [11]. This study demonstrated that resveratrol acts as an antioxidant and antimutagen and induces phase II drug-metabolizing enzymes (anti-initiation activity). In addition, resveratrol was found to mediate anti-inflammatory effects and inhibited cyclooxygenase and hydroperoxidase functions (antipromotion activity), and it also induced the differentiation of human promyelocytic leukemia cells (antiprogression activity). These data, for the first time, suggested that resveratrol merits an investigation as a potential cancer chemopreventive agent [11].

Another recent study suggested that resveratrol inhibits tumorigenesis in mouse skin through interference with pathways of reactive oxidants and possibly by modulating the expression of c-*fos* and TGF-β_1 [12]. In this study, the application of TPA to mouse skin resulted in (1) significant generation of H_2O_2, (2) enhanced levels of myeloperoxidase and oxidized glutathione (GSH) reductase activities and (3) decreases in GSH levels and superoxide dismutase activity. A pretreatment of skin with resveratrol resulted in reversal of these effects [12]. In addition, resveratrol treatment also resulted in an inhibition of the TPA-mediated increase in the expression of cyclooxygenases (COX-1 and COX-2), c-*myc*, c-*fos*, c-*jun*, TGF-β_1 and TNF-α [12].

Skin Cancer Chemoprevention by Ginger Compounds

Studies have shown the anti-skin-cancer chemopreventive properties of ginger [13]. Katiyar et al. [13] demonstrated that a topical preapplication of an ethanol extract of ginger on the skin of SENCAR mice resulted in a significant inhibition of TPA-caused induction of epidermal ornithine decarboxylase (ODC), cyclooxygenase and lipoxygenase activities and ODC mRNA expression. In this study, ginger extract was also found to inhibit TPA-caused epidermal edema and hyperplasia in SENCAR mice. Further, in the long-term tumor studies, topically applied ginger extract resulted in a highly significant protection against DMBA-initiated TPA-promoted skin tumor incidence and multiplicity in SENCAR mice [13]. This study, for the first time, demonstrated that ginger possesses anti-skin-cancer properties.

In a subsequent study, Park et al. [14] evaluated the antitumor promotional activity of 6-gingerol, the major pungent constituent of ginger, in a two-stage mouse skin carcinogenesis model. The topical application of 6-gingerol on the dorsal shaven skin of female ICR mice significantly inhibited DMBA-initiated TPA-promoted skin papilloma formation, as well as TPA-induced epidermal ODC activity and inflammation in mouse skin [14]. The cancer chemopreventive properties of ginger have been attributed to the antioxidant potential of the constituents present therein. Based on many studies, it is believed that gingerol is a good scavenger of peroxyl radicals generated by pulse radiolysis.

Skin Cancer Chemoprevention by Curcumin

Curcumin, a yellow ingredient from turmeric (*Curcuma longa*), has been extensively investigated for its cancer chemopreventive potential [reviewed in 15]. Studies have shown that it exhibits antimutagenic activity in the Ames Salmonella test and possessed anticarcinogenic activity as it inhibits chemically induced neoplastic lesions in many organs including the skin, probably via an antioxidant mechanism [reviewed in 15]. Curcumin has been shown to enhance GSH content and glutathione-S-transferase activity and inhibits lipid peroxidation and arachidonic acid metabolism in mouse skin [reviewed in 15]. The antioxidant and anti-inflammatory properties of curcumin have been well documented [reviewed in 15].

Curcumin has been shown to inhibit DMBA-initiated, TPA-promoted skin tumors in female Swiss mice, suggesting that curcumin inhibits cancer at initiation, promotion and progression stages of development [reviewed in 15]. Huang et al. [16] demonstrated that a topical application of curcumin resulted in inhibition of TPA-mediated induction of epidermal ODC activity, stimulation of the incorporation of [^{3}H]thymidine into epidermal DNA and epidermal DNA synthesis in female CD-1 mice. This treatment also resulted in a significant inhibition of DMBA-initiated TPA-promoted skin tumorigenesis in these mice [reviewed in 16].

In another study by Huang et al. [reviewed in 16], the effects of topical application of curcumin on the formation of benzo[a]pyrene (B[a]P) DNA adducts and the tumorigenic activities of B[a]P and DMBA were evaluated in skin of female CD-1 mice. In this study, curcumin treatment resulted in an inhibition in the formation of [^{3}H]B[a]P DNA adducts in epidermis [reviewed in 16]. In a B[a]P TPA model of skin tumorigenesis, a topical application of curcumin prior to B[a]P application resulted in a decrease in the number of tumors per mouse and the percentage of tumor-bearing mice [reviewed in 16].

Nakamura et al. [17] investigated the inhibitory effects of curcumin and its derivatives on tumor-promoter-induced oxidative stress in vitro and in vivo. Curcumin, tetrahydrocurcumin and dihydroxytetrahydrocurcumin were found to impart significant inhibitory effects on TPA-induced oxygen generation in differentiated HL-60 cells. The curcuminoids also inhibited TPA-induced intracellular peroxide formation in these cells [17]. This study further examined the effects of curcuminoids on TPA-induced H_2O_2 formation in female ICR mouse skin using the double-TPA-application model [17]. In this model, each TPA application induced two distinct biochemical events, i.e. the recruitment of inflammatory cells to the inflammatory regions and activation of oxidant-producing cells. The data from this study revealed that double pretreatment of mice with curcuminoids prior to each TPA treatment significantly reduced TPA-induced H_2O_2 formation in the mouse skin [17]. This study suggested that curcumin and its related compounds inhibit skin tumorigenesis via an antioxidant pathway.

Ishizaki et al. [18] demonstrated that UVA irradiation (18.72 J/cm^2) significantly enhanced ODC induction after topical application of TPA in the epidermis of CD-1 mice, and aggravated TPA-mediated dermatitis. A pretreatment of skin with curcumin was found to significantly inhibit these enhancing effects [18]. This study suggested that curcumin may impart a beneficial effect against the responses of UV radiation in skin.

Skin Cancer Chemoprevention by Diallyl Sulfide

Diallyl sulfide, a naturally occurring compound present in garlic and onion, is known to possess strong antioxidant potential [19 and references therein]. Perchellet et al. [20] have shown that these agents increase GSH peroxidase (GSH: H_2O_2 oxidoreductase) activity in isolated epidermal cells incubated in the presence or absence of TPA. The stimulatory effects of these oils on epidermal GSH peroxidase activity were found to be concentration-dependent and long-lasting. These oils completely abolished the prolonged inhibitory effect of TPA on this enzyme [20]. Further, garlic oil significantly inhibited TPA-induced ODC activity in the same epidermal cell system and enhanced GSH peroxidase activity in the presence of various nonphorbol ester tumor promoters [20]. Based on this study, it was suggested that some of the inhibitory effects of garlic and onion oils on skin tumor promotion may result from their enhancement of the natural GSH-dependent antioxidant protective system of the epidermal cells.

Studies have implicated diallyl sulfide as a chemopreventive agent against a variety of cancer types including skin cancer. Sadhana et al. [21] demonstrated that topical application of garlic oil resulted in a significant inhibition of benzo[a]pyrene-induced skin tumor formation in female Swiss albino mice.

Diallyl sulfide treatment was found to result in a decrease in the number of tumor-bearing mice and in the mean number of tumors per mouse [21]. In another study, Perchellet et al. [22] showed that garlic oil and onion oils inhibited skin tumor promotion in SENCAR mice. This study further revealed that garlic oil inhibited DMBA-induced mouse skin tumorigenesis [22]. In another study from this laboratory, Athar et al. [19] demonstrated that topical application of diallyl sulfide resulted in an inhibition of benzoyl-peroxide-mediated tumor promotion in DMBA-initiated skin cancer of SENCAR mice.

In a study by Dwivedi et al. [23], it was found that topical applications of diallyl sulfide and diallyl disulfide to the skin of SENCAR mice resulted in significant inhibition of DMBA-induced and TPA-promoted skin tumor formation. Singh and Shukla [24] demonstrated that diallyl sulfide inhibits B[a]P- and DMBA-induced carcinogenesis.

Skin Cancer Chemoprevention by Silymarin

Silymarin, a flavonoid extracted from the seeds of *Silybum marianum*, is a mixture of at least three structural isomers: silybin, silydianin and silychristin, the former being the most active component. Silymarin is known to be an antioxidant compound with skin cancer chemopreventive properties [reviewed in 6]. In a study from this laboratory, Agarwal et al. [reviewed in 6] assessed the effect of the topical application of silymarin on TPA-induced epidermal ODC activity and ODC mRNA levels in SENCAR mouse skin. Silymarin treatment resulted in a significant inhibition of TPA-induced ODC activity and ODC mRNA expression in the epidermis. In this study, silymarin also showed significant inhibition of epidermal ODC activity induced by several other tumor promoters and free-radical-generating compounds [reviewed in 6]. These data suggested that silymarin could be a useful chemopreventive agent for skin cancer.

In a subsequent study from this laboratory, Chatterjee et al. [25], employing a ^{32}P post-labeling technique, demonstrated that topical application of silymarin or GTP as well as a sunscreen containing ethylhexyl-p-methoxy-cinnamate resulted in a protection against UVB-exposure-mediated formation of pyrimidine dimers in mouse skin.

Katiyar et al. [26] evaluated the protective effects of silymarin against UVB-radiation-induced nonmelanoma skin cancer in mice in long-term and short-term studies. In this study, female Skh-1 hairless mice were subjected to UVB-induced tumor initiation followed by TPA-mediated tumor promotion, DMBA-induced tumor initiation followed by UVB-mediated tumor promotion and UVB-induced complete carcinogenesis. Silymarin was topically applied prior to UVB exposure, and its effects on tumor incidence (percentage of mice with tumors), tumor multiplicity (number of tumors per mouse) and average tumor volume per mouse were recorded [26]. This study demonstrated that in all the protocols employed, silymarin treatment was found to result in significant protection against skin tumorigenesis in the mice [26]. Further, in short-term experiments, silymarin application was found to result in significant inhibition of UVB-induced formation of sunburn and apoptotic cells, skin edema, depletion of catalase activity and induction of COX and ODC activities and ODC mRNA expression [26]. This study suggested that silymarin may provide protection against different stages of UVB-induced carcinogenesis, possibly via its antioxidant properties.

In a recent study, Lahiri-Chatterjee et al. [27] assessed the protective effect of silymarin on tumor promotion in the SENCAR mouse skin tumorigenesis model. In this study it was demonstrated that topical preapplication of silymarin resulted in a highly significant protection against DMBA/TPA-mediated skin carcinogenesis [27]. To evaluate the stage specificity of silymarin against tumor promotion, this study assessed the effect of silymarin in both stage I and stage II of tumor promotion. It was found that topical application of silymarin prior to that of TPA in stage I or mezerein in stage II tumor promotion in DMBA-initiated SENCAR mouse skin resulted in a significant protective effect in both protocols [27]. In this study, silymarin also inhibited: TPA-induced skin edema, epidermal hyperplasia and proliferating cell-nuclear-antigen-positive cells, DNA synthesis, and epidermal lipid peroxidation [27]. These results suggested that silymarin possesses protective effects against chemically induced skin carcinogenesis.

Zhao et al. [28] demonstrated that topical application of silymarin onto SENCAR mouse skin resulted in inhibition of TPA-mediated skin edema and depletion of epidermal superoxide dismutase, catalase and glutathione peroxidase activities. Silymarin also inhibited TPA-mediated epidermal lipid peroxidation and myeloperoxidase activity [28].

Conclusion

Skin cancer is a potential problem associated with significant mortality and morbidity in the human population. Skin cancer is estimated to attack 1 out of every 7 Americans each year, making it the most prevalent form of cancer. More than a million new cases of nonmelanoma skin cancers are diagnosed annually in the USA. Another disturbing fact about this cancer is that there is an increased risk for the individuals with a history of skin cancer towards other lethal cancer types. In view of these facts, there is an urgent need to develop mechanism-based approaches for prevention/therapy of skin cancer. Chemoprevention by naturally occurring antioxidant compounds is such an approach. In recent times, the concept of chemoprevention is being increasingly accepted. This can also be appreciated by the fact that, at present, a variety of cosmetic products supplemented with green tea are available at the cosmetic counters of drug stores, supermarkets and department stores. Products supplemented with botanicals such as green tea include, but are not limited to, depilatory creams, cleansing lotions, shampoos, moisturizing creams, toothpastes, scented sprays, body lotions, etc. Most of the naturally occurring antioxidant compounds discussed in this review have not been adequately tested for their effectiveness and safety for humans in clinical trials. Therefore, there is a need to initiate clinical trials for selected antioxidants for skin cancer chemoprevention.

References

1 American Cancer Society-Cancer Facts and Figures: Graphical data (available at the web site http://www.cancer.org).
2 Kahn HS, Tatham LM, Patel AV, Thun MJ, Heath CW: Increased cancer mortality following a history of non melanoma skin cancer. JAMA 1998;280:910–912.
3 Steenvoorden DP, van Henegouwen GM: The use of endogenous antioxidants to improve photoprotection. J Photochem Photobiol B 1997;41:1–10.
4 Mukhtar H, Ahmad N: Cancer chemoprevention: Future holds in multiple agents: Contemporary issues in toxicology. Toxicol Appl Pharmacol 1999;158:207–210.
5 Mukhtar H, Mercurio MG, Agarwal R: Murine skin carcinogenesis: Relevance to humans; in Mukhtar H (ed): Skin Cancer: Mechanism and Human Relevance. Boca Raton, CRC Press, 1995, pp 3–8.
6 Mukhtar H, Agarwal R: Skin cancer chemoprevention. J Invest Dermatol Symp Proc 1996;1: 209–214.
7 Katiyar SK, Mukhtar H: Tea antioxidants in cancer chemoprevention. J Cell Biochem 1997; 27(suppl):59–67.
8 Rice-Evans C: Implications of the mechanisms of action of tea polyphenols as antioxidants in vitro for chemoprevention in humans. Proc Soc Exp Biol Med 1999;220:262–266.
9 Katiyar SK, Matsui MS, Elmets CA, Mukhtar H: Polyphenolic antioxidant (–)-epigallocatechin-3-gallate from green tea reduces UVB-induced inflammatory responses and infiltration of leukocytes in human skin. Photochem Photobiol 1999;69:148–153.
10 Katiyar SK, Ahmad N, Mukhtar H: Green tea and skin. Arch Dermatol 2000;136:989–994.
11 Jang M, Cai L, Udeani GO, Slowing KV, Thomas CF, Beecher CW, Fong HH, Farnsworth NR, Kinghorn AD, Mehta RG, Moon RC, Pezzuto JM: Cancer chemopreventive activity of resveratrol, a natural product derived from grapes. Science 1997;275:218–220.
12 Jang M, Pezzuto JM: Effects of resveratrol on 12-O-tetradecanoylphorbol-13–acetate-induced oxidative events and gene expression in mouse skin. Cancer Lett 1998;134:81–89.
13 Katiyar SK, Agarwal R, Mukhtar H: Inhibition of tumor promotion in SENCAR mouse skin by ethanol extract of Zingiber officinale rhizome. Cancer Res 1996;56:1023–1030.
14 Park KK, Chun KS, Lee JM, Lee SS, Surh YJ: Inhibitory effects of 6-gingerol, a major pungent principle of ginger, on phorbol ester-induced inflammation, epidermal ornithine decarboxylase activity and skin tumor promotion in ICR mice. Cancer Lett 1998;129:139–144.
15 Stoner GD, Mukhtar H: Polyphenols as cancer chemopreventive agents. J Cell Biochem 1995; 22(suppl):169–180.
16 Huang MT, Newmark HL, Frenkel K: Inhibitory effects of curcumin on tumorigenesis in mice. J Cell Biochem 1997;27(suppl):26–34.

17 Nakamura Y, Ohto Y, Murakami A, Osawa T, Ohigashi H: Inhibitory effects of curcumin and tetrahydrocurcuminoids on the tumor promoter-induced reactive oxygen species generation in leukocytes in vitro and in vivo. Jpn J Cancer Res 1998;89:361–370.
18 Ishizaki C, Oguro T, Yoshida T, Wen CQ, Sueki H, Iijima M: Enhancing effect of ultraviolet A on ornithine decarboxylase induction and dermatitis evoked by 12-O-tetradecanoylphorbol-13-acetate and its inhibition by curcumin in mouse skin. Dermatology 1996;193:311–317.
19 Athar M, Raza H, Bickers DR, Mukhtar H: Inhibition of benzoyl peroxide-mediated tumor promotion in 7,12-dimethylbenz(a)anthracene-initiated skin of Sencar mice by antioxidants nordihydroguaiaretic acid and diallyl sulfide. J Invest Dermatol. 1990;94:162–165.
20 Perchellet JP, Perchellet EM, Abney NL, Zirnstein JA, Belman S: Effects of garlic and onion oils on glutathione peroxidase activity, the ratio of reduced/oxidized glutathione and ornithine decarboxylase induction in isolated mouse epidermal cells treated with tumor promoters. Cancer Biochem Biophys 1986;8:299–312.
21 Sadhana AS, Rao AR, Kucheria K, Bijani V: Inhibitory action of garlic oil on the initiation of benzo[a]pyrene-induced skin carcinogenesis in mice. Cancer Lett 1988;40:193–197.
22 Perchellet JP, Perchellet EM, Belman S: Inhibition of DMBA-induced mouse skin tumorigenesis by garlic oil and inhibition of two tumor-promotion stages by garlic and onion oils. Nutr Cancer 1990;14:183–193.
23 Dwivedi C, Rohlfs S, Jarvis D, Engineer FN: Chemoprevention of chemically induced skin tumor development by diallyl sulfide and diallyl disulfide. Pharm Res 1992;9:1668–1670.
24 Singh A, Shukla Y: Antitumour activity of diallyl sulfide on polycyclic aromatic hydrocarbon-induced mouse skin carcinogenesis. Cancer Lett 1998;131:209–214.
25 Chatterjee ML, Agarwal R, Mukhtar H: Ultraviolet B radiation-induced DNA lesions in mouse epidermis: An assessment using a novel ^{32}P-postlabelling technique. Biochem Biophys Res Commun 1996;229:590–595.
26 Katiyar SK, Korman NJ, Mukhtar H, Agarwal R: Protective effects of silymarin against photocarcinogenesis in a mouse skin model. J Natl Cancer Inst 1997;89:556–566.
27 Lahiri-Chatterjee M, Katiyar SK, Mohan RR, Agarwal R: A flavonoid antioxidant, silymarin, affords exceptionally high protection against tumor promotion in the SENCAR mouse skin tumorigenesis model. Cancer Res 1999;59:622–632.
28 Zhao J, Sharma Y, Agarwal R: Significant inhibition by the flavonoid antioxidant silymarin against 12-O-tetradecanoylphorbol 13-acetate-caused modulation of antioxidant and inflammatory enzymes, and cyclooxygenase 2 and interleukin-1alpha expression in SENCAR mouse epidermis: Implications in the prevention of stage I tumor promotion. Mol Carcinog 1999;26:321–333.

Hasan Mukhtar, PhD, Department of Dermatology, Case Western Reserve University,
11100 Euclid Avenue, Cleveland, OH 44106 (USA)
Tel. +1 (216) 368 1127, Fax +1 (216) 368 0212, E-Mail hxm4@po.cwru.edu

Thiele J, Elsner P (eds): Oxidants and Antioxidants in Cutaneous Biology.
Curr Probl Dermatol. Basel, Karger, 2001, vol 29, pp 140–156

Radical Reactions of Carotenoids and Potential Influence on UV Carcinogenesis

Homer S. Black[a], *Christopher R. Lambert*[b]

[a] Baylor College of Medicine and Veterans Affairs Medical Center, Houston, Tex., and
[b] Department of Chemistry, Connecticut College, New London, Conn., USA

The carotenoids are a class of linear polyenes that are found in plants, algae and in some bacteria and fungi (fig. 1). Strictly speaking, carotenoids containing oxygen, such as canthaxanthin and astaxanthin, form a subclass called the xanthophylls. Around 600 carotenoids have been identified thus far [1]. Chemically, typical carotenoid pigments are tetraterpenoids, consisting of 8 isoprenoid residues. The carotenoids exhibit a very strong light absorption in the region of 400–500 nm and thus are colored yellow, orange or red, depending on concentration. Sistrom et al. [2] and Mathews-Roth and Sistrom [3] first showed that these pigments protect photosynthetic organisms against potentially lethal photosensitization by endogenous photosensitizers such as the porphyrin-containing photosynthetic pigments. In 1951, Kesten [4] reported that the oral intake of the carotenoid, β-carotene, delayed erythema onset in a patient suffering from severe photosensitivity. Armed with the knowledge of the carotenoid's protective role in porphyrin-related photosensitization and the presumptive evidence of Kesten, Mathews-Roth theorized that carotenoid pigments might serve a protective role in humans similar to that in plants [5, 6]. In a practical application related to these observations, Mathews [7] demonstrated that β-carotene effectively prevented lethal hematoporphyrin photosensitization in mice and then employed the pigment in the treatment of protoporphyrin-induced photosensitivity in the human genetic disease erythropoietic protoporphyria [8].

Of the approximate 600 carotenoids identified, about 100 occur in foods eaten by humans [1]. The carotenoid pigments are widely distributed as natur-

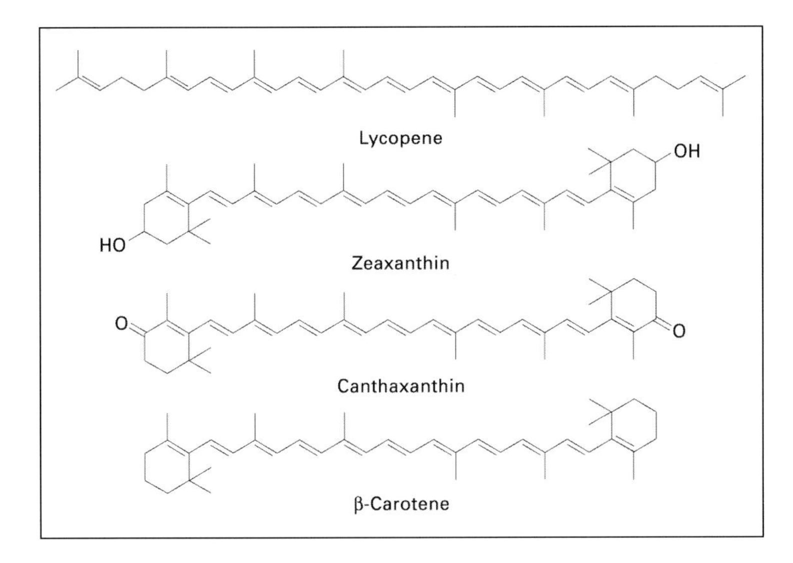

Fig. 1. Carotenoid structures.

ally occurring constituents of fruits and vegetables, especially green leafy and yellow vegetables. In addition, carotenoids find their way into the human food chain as additives to foodstuffs in order to achieve acceptable food coloration [9]. The estimated daily consumption of β-carotene for adults in the USA ranges from $\leq$2.0 mg in the lowest quartile to an average of 6.0 mg in the highest quartile [10]. These values were based upon a carotene index calculated from the amount of carotene per unit of food × units consumed. The characteristic carotene value of the food was obtained from a nutritional database.

In 1981, Peto et al. [11] reported that human cancer risks were inversely correlated with β-carotene intake. The study was based upon the consumption of foodstuffs rich in β-carotene and stimulated interest in the carotenoids, especially β-carotene, as cancer-preventive agents. A large number of epidemiological studies have followed that examined the relationship between carotenoid-rich food intake and risks of certain types of cancer [1, 12]. The associations have been strongest with lung cancer with almost all studies showing an inverse relationship between blood β-carotene levels or total carotene status and lung cancer risk [1]. A case-control study also found that the incidence of nonmelanoma skin cancer was inversely related to the level of serum β-carotene [13]. With regard to nonmelanoma skin cancer, a series of experimental UV carcinogenesis studies conducted in the 1970s and 1980s also found β-carotene to be photoprotective [14–17].

However, the role of β-carotene as an anticancer agent has recently been questioned as a result of intervention clinical studies in which the incidence of nonmelanoma skin cancer was unchanged in patients receiving a β-carotene supplement and in β-carotene-supplemented smokers who suffered a significant increase in lung cancer occurrence [18, 19]. Further, in recent laboratory studies, β-carotene-supplemented semidefined diets, in contrast to earlier studies employing commercial closed-formula diets, not only failed to provide a protective effect against UV carcinogenesis but resulted in significant *exacerbation* [20]. Because these results, as well as those of the smokers' clinical trial, point to possible risks of single carotenoid dietary supplementation and because the rationale for this distinct carcinogenic response to β-carotene may rest with the carotenoid's specific capacity to quench singlet oxygen and interact with other oxy- and carbon-centered radicals, it is important to examine the mechanism(s) by which these reactions proceed if we are to understand the anti/pro-cancer responses to β-carotene.

Radical Reactions of Carotenoids

Carotenoids are efficient quenchers of singlet oxygen [21]. The rate of quenching has been found to correlate with the length of the carotenoid's conjugated polyene chain and thought to involve electron exchange energy transfer [22], i.e.

$$O_2(^1\Delta_g) + Car \rightarrow {}^3Car + O_2. \quad (1)$$

The reverse process does not lead to the generation of singlet oxygen. The rate constant for quenching of singlet oxygen by carotenoids is typically close to diffusion controlled and it is difficult to correlate this physicochemical parameter with any protective effect, although it was generally believed that the rate of singlet oxygen quenching by a carotenoid paralleled its protective action [23]. However, in a biological system, Mathews-Roth et al. [24] found that the protective action of a series of carotenoid pigments did not necessarily parallel the $O_2(^1\Delta_g)$ quenching capacity and suggested that carotenoids might interfere with radical reactions initiated in vivo. Subsequently, it was shown that carotenoids could effectively inhibit lipid peroxidation in microsomal membranes by mechanisms not initiated by $O_2(^1\Delta_g)$, their antioxidant efficiency influenced by several factors including the type of radical initiator involved and the site and rate of radical formation [25, 26]. Furthermore, in work that may be pivotal to our understanding of the biological responses to carotenoids, Burton and Ingold [27] showed that β-carotene exhibited good radical-trapping antioxidant behavior at partial pressures of O_2 significantly less than 150 Torr (pressure of O_2 in normal air). At higher O_2 pressure β-carotene lost its antioxidant capacity and showed autocatalytic, prooxidant effects.

β-Carotene is highly reactive toward some model peroxy radicals, e.g. the trichloromethyl peroxyl radical, $CCl_3O_2^{\bullet}$ [28], forming a carotenoid radical cation by positive charge transfer from solvent cations:

$$CCl_3O_2^{\bullet} + Car \rightarrow CCl_3O^{-} + Car^{\bullet +}. \quad (2)$$

Indeed, of a list of carotenoids investigated, including β-carotene, septapreno-β-carotene, canthaxanthin, astaxanthin. zeaxanthin and lutein, all reacted with $CCl_3O_2^{\bullet}$ to produce two reaction products, an addition radical and the carotenoid radical cation [28–30]. Thus, with an oxygen-centered radical, such as the model peroxyl radical ($RO_2^{\bullet}$), reactions may proceed either via electron transfer to produce the carotenoid radical cation

$$RO_2^{\bullet} + CAR \rightarrow RO_2^{-} + CAR^{\bullet +} \quad (3)$$

or by other processes such as addition reaction to a carbon-carbon double bond,

$$RO_2^{\bullet} + Car \rightarrow [RO_2^{-} - Car]^{\bullet}, \quad (4)$$

or hydrogen atom transfer:

$$RO_2^{\bullet} + Car \rightarrow ROOH + Car(-H^{\bullet}). \quad (5)$$

Nevertheless, while B-carotene is highly reactive toward peroxyl radicals and is a good agent for reducing the level of chain-carrying peroxyl radicals at low oxygen partial pressure, its antioxidant activity depends upon the production of a resonance-stabilized carbon-centered radical. At higher oxygen partial pressures, β-carotene appears to act as a prooxidant, reacting minimally with model compounds such as methyl linoleate [27]. Interestingly, in a prelude to our understanding of the physiological interactions of vitamin E (α-tocopherol) and β-carotene, these authors reported that the radical chain-breaking action of vitamin E was most effective at high oxygen concentrations, a condition where the prooxidant activity of β-carotene would most likely occur. In this regard, Anderson et al. [31], using Fe^{2+}-mediated oxidation of heme proteins as an early indicator of oxidative stress, found a pronounced vitamin E antioxidant activity whereas carotenoids exhibited either limited protection or a prooxidation effect.

Truscott [32] and colleagues found that at very low oxygen concentrations only the product of equation 3 was observed, i.e. electron transfer producing the carotenoid radical cation. They also found a good correlation between the rate of decay of the second species found above, i.e. [RO2–β-Car]$^{\bullet}$, and growth of the β-carotene radical cation. Based upon these observations, Truscott proposed a mechanism for the interaction of β-carotene and peroxyl radicals and oxygen to explain the carotenoid's antioxidant action at low oxygen concentrations and the prooxidant behavior at high oxygen levels. The major antioxidant route entails removal of the chain initiating peroxyl species by β-carotene to produce the carotenoid radical cation. The latter is rather long lived, does not significantly react with oxygen and would not propagate the chain reaction,

$$RO_2 + \beta\text{-Car} \rightarrow [RO_2 - \beta\text{-Car}]^{\bullet} \rightarrow \beta\text{-Car}^{\bullet+}. \quad (6)$$

However, under high oxygen concentrations the prooxidant mechanism would involve the formation of an oxygenated β-carotene product and a radical that could reform the peroxyl radical and continue the chain reaction:

$$[RO_2 - \beta\text{-Car}]^{\bullet} \rightarrow \beta\text{-Car-}O_2 + R^{\bullet} \text{ and} \quad (7)$$

$$R^{\bullet} + O_2 \rightarrow RO_2^{\bullet}. \quad (8)$$

The oxygenated β-carotene product ($[RO_2\text{–}\beta\text{-Car}]^{\bullet}$) might then react with membrane lipid (LH) to form lipid hydroperoxides (LOOH) or other oxidation products:

$$LH + \beta\text{-Car–}O_2 \rightarrow LOOH + \beta\text{-Car}. \quad (9)$$

Although this mechanism to explain both prooxidant and antioxidant behavior of β-carotene remains speculative, it is consistent with the reactions that have been observed, albeit in homogenous solvents and micelles. Thus, localization and orientation of a carotenoid within a biological membrane may have marked limitations upon its capacity for reaction.

As noted in the antioxidant mechanism for terminating the peroxyl chain reaction above (equation 6), the β-carotene radical cation is the final product. Although this radical does not continue the peroxyl chain reaction, it is relatively long lived. This leads one to question whether these rather persistent radicals might directly contribute to membrane damage or indirectly to the prooxidant activity of the parent compound. The former would depend upon whether the carotenoid radical cation could be 'repaired'. To address this question, Edge et al. [33] have examined the electron transfer rate constants between various pairs of carotenoids and their radical cations. Based upon the carotenoids studied, they were placed in the following relative order in terms of reduction potential: astaxanthin > β-apo-8′-carotenal > canthaxanthin > lutein > zeaxanthin > β-carotene > lycopene. These data suggest that lycopene efficiently quenches (repairs) most carotenoid radical cations, e.g.

$$Car^{\bullet+} + Lyc \rightarrow Car + Lyc^{\bullet+}. \quad (10)$$

Based on these studies [33] and studies of the interactions of carotenoids with vitamin E (tocopherol, TOH) and vitamin C (ascorbic acid, $AscH^-$),

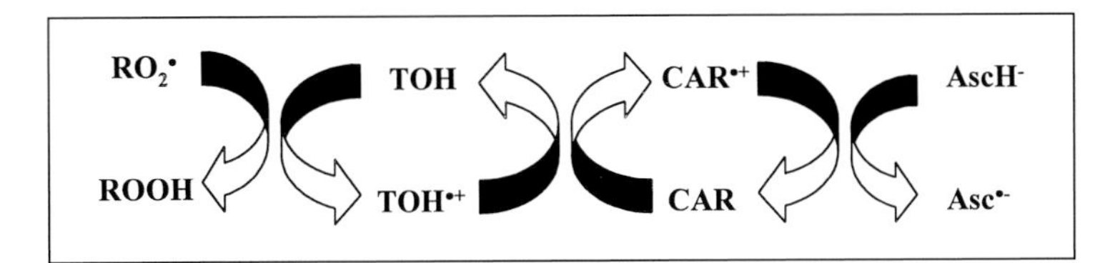

Fig. 2. Proposed mechanism by which β-carotene participates in quenching oxy radicals and interacts to enhance the antioxidant properties of vitamin E. Based upon the relative electron transfer rate constants for interactions between β-carotene, vitamin E (α-tocopherol, TOH) and vitamin C (ascorbic acid, $AscH^-$), α-tocopherol would first intercept an oxy radical. In terminating the radical-propagating reaction, the tocopherol radical cation ($TOH^{\bullet +}$) is formed which, in turn, is repaired by β-carotene to form the carotenoid radical cation ($CAR^{\bullet +}$). This radical is repaired by ascorbic acid. Because of its hydrophilic characteristics, it is anticipated that the ascorbic acid radical would be formed in the hydration shell surrounding the membrane and either cleared or enzymatically repaired before it induced membrane damage.

Bohm et al. [34] suggested a mechanism for the interaction of β-carotene with vitamin E and C radicals which ultimately resulted in the repair of the β-carotene radical cation. In this scenario, developed from previous studies of the interaction of vitamins C and E [35], β-carotene repairs the tocopherol radical to produce the carotenoid radical cation which, in turn, is repaired by vitamin C. This interpretation of the data [28, 32–34] presented an interesting problem. Valgimigli et al. [36], employing electron paramagnetic resonance spectroscopy, found no evidence for the reaction of the α-tocopherol radical ($TOH^{\bullet}$) with β-carotene. Electron paramagnetic resonance spectroscopy has the advantage over spectroscopically monitored pulse radiolysis in that radical species can be unequivocally identified. More recently, Edge et al. [33] have shown that β-carotene does react with the α-tocopherol radical cation ($TOH^{\bullet +}$) to produce the carotenoid radical cation. This radical can be repaired by ascorbic acid, i.e.

$$TOH^{\bullet +} + \beta\text{-Car} \rightarrow TOH + Car^{\bullet +} \text{ and} \tag{11}$$

$$Car^{\bullet +} + AscH^{-} \rightarrow Car + Asc^{\bullet} + H^{+}. \tag{12}$$

The schema shown in figure 2, based on equations 11 and 12, has been proposed as a mechanism by which β-carotene not only quenches oxy radicals but could enhance the radical-protective properties of both vitamins E and C, as well.

The difficulties in understanding the antioxidant characteristics of β-carotene, exemplified by the different anti/prooxidant responses observed under various experimental conditions (e.g. in homogenous solvents, micelles and

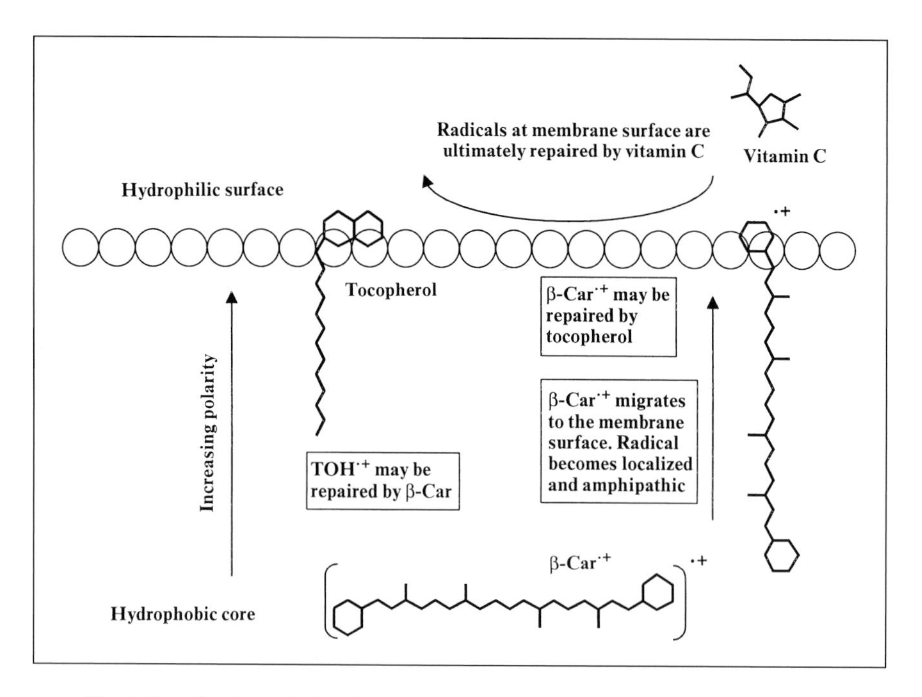

Fig. 3. Localization of vitamin E and β-carotene within a membrane and a proposal for carotenoid reactivity. The β-Car$^{\bullet+}$ may form either as a result of reaction with a lipid peroxyl radical or TOH$^{\bullet+}$. β-Car$^{\bullet+}$ migrates to the surface of the membrane, and the radical center becomes localized at one end of the carotenoid molecule. β-Car$^{\bullet+}$ may be repaired by vitamin E or C. The diagram is stylistic and not to scale.

in vivo) and synergies of antioxidant equivalents between water- and lipid-soluble antioxidants, illustrate the complexities that must be considered [37].

That β-carotene can react with TOH$^{\bullet+}$ but not TOH$^{\bullet}$ leads to the unavoidable conclusion that the reduction potential of β-carotene must lie between that of TOH$^{\bullet+}$ and TOH$^{\bullet}$. This suggests that antioxidant reactions within the membrane are significantly more complex than previously thought. Figure 3 attempts to summarize the important aspects of the vitamin E/β-carotene interaction.

The overall mechanism is complicated because radicals at the membrane surface may deprotonate and the reduction potential is dependent on both the 'solvent' and the protonation state. The Car$^{\bullet+}$ at the membrane surface is also likely to be more reactive since the unpaired electron is no longer delocalized over the entire carotenoid molecule. β-Car$^{\bullet+}$ may react with vitamin E or vitamin C. The membrane-based radical, whether a vitamin E radical or β-Car$^{\bullet+}$, is ultimately removed by reaction with water-soluble vitamin C.

Reaction of Trolox (a water-soluble vitamin E derivative) and β-Car$^{\bullet+}$ has been observed by pulse radiolysis [Ruth Edge, pers. commun.].

In addition to the mechanistic concerns, under in vivo conditions, absorption, target tissue concentrations, rate constants for radical reactions (unfortunately the direct kinetic experiments cannot currently be conducted in vivo [37]), localization and mobility with respect to hydrophobic and hydrophilic domains, turnover rates of the antioxidant in the respective tissue and rate of regeneration or recycling are but some of the factors that impact an antioxidant's efficacy. Nevertheless, the mechanism proposed in figure 2 represents a rational model that can explain how the carotenoid radical cation is repaired and that could account for the contradictions of β-carotene's antioxidant properties that have arisen from epidemiological, intervention and experimental studies.

Epidemiological, Intervention and Experimental Studies

Almost all of the large number of prospective and retrospective epidemiological studies of either the intake of foods rich in β-carotene or high levels of blood β-carotene have found a strong association with reduced risks of various kinds of cancer. Shekelle et al. [10] had reported, in 1981, findings of a 19-year longitudinal case-control study in which they found that smokers had a markedly greater incidence of lung cancer than nonsmokers and that the greatest incidence occurred in the quartile that exhibited the lowest carotene intake. In fact, of nine major case-controlled studies of lung cancer and carotene intake, nearly all showed a lower risk of lung cancer among people with a higher intake of carotenoids [1]. In addition, seven nested case-control studies of lung cancer and blood (serum or plasma) β-carotene concentrations all found significant inverse associations when controlled for tobacco use and other factors. However, a majority found stronger inverse trends with vegetable and fruit intake than with the estimated carotenoid intake. Thus, there was a hint that the combination of β-carotene with other micronutrients may have had a greater effect than the carotenoid alone.

In the face of overwhelming epidemiological evidence for a cancer-preventive effect of β-carotene, it was surprising, and disturbing, when results from the 8-year intervention trial of the Alpha-Tocopherol, Beta-Carotene Cancer Prevention Study found an 18% increase in incidence of lung cancer in smokers who were given a 20 mg/day β-carotene supplement, compared to nonsupplemented controls [19]. Nor is this the only case where β-carotene has failed to provide beneficial effects in intervention trials. In fact, of the five intervention trials undertaken, none, except a trial where β-carotene was administered in combination with selenium and α-tocopherol, provided evidence of a protective effect of β-carotene [1].

One of the aforementioned studies, the Skin Cancer Prevention Study, examined the effect of β-carotene supplementation (50 mg/day) on the occurrence of subsequent nonmelanoma skin cancers in patients who had been treated for a previous skin cancer. This study also examined the effectiveness of β-carotene supplementation in reducing the average number of new skin cancers per person [18]. A previous case-control study had shown a significant inverse trend between serum β-carotene concentration and skin cancer risk [13]. However, after a 5-year treatment period in the intervention trial, β-carotene supplementation at a level that increased the plasma concentration about 10-fold, had no effect overall on any of the predefined primary endpoints. Subsequent evaluations from nested case-control studies showed that β-carotene supplementation had no effect in any of the controlled patient subgroups, i.e. numbers of previous skin cancers, age, gender, smoking, skin type or baseline plasma β-carotene levels [38, 39]. Interestingly, the investigators found that those persons in the study who were in the highest quartile of the initial plasma β-carotene level had a lower risk of death from all causes. However, β-carotene supplementation did not affect mortality.

There have been few studies that have examined the influence of β-carotene on the occurrence of melanoma skin cancer. However, in three case-control studies, no association was found between blood carotenoid levels and the reduced risk of melanoma [40–42]. In one of the studies, risk among men and women in the three highest quartiles of β-carotene intake *increased* by 40–50%, albeit neither the risk for the fourth quartile nor the test for trend was statistically significant [42]. It is interesting to note that β-carotene supplementation (50 mg/day for 36 months) has been reported to have beneficial effects in dysplastic nevus syndrome, producing a statistically significant retardation of increase in size of dysplastic nevi on some body sites, but not on the average of newly developed nevi [43].

In 1972, Mathews-Roth et al. [44] demonstrated that β-carotene exerted a small, but statistically significant, effect in increasing the minimal erythema dose to sunburn in man. Studies were extended to show a similar photoprotective effect with phytoene in guinea pigs [45]. Phytoene is the triene precursor to β-carotene and absorbs strongly in the UVB range of the spectrum. Based upon these observations, Epstein [14] examined the potential influence of intraperitoneally injected β-carotene on UV-induced tumor formation in the hairless mouse. β-Carotene was found to exert a limited, but protective, effect both in regard to time of tumor appearance and tumor growth. In mixed chemical/UV carcinogenic protocols, oral administrations of β-carotene or canthaxanthin were found to delay the appearance of skin tumors [15, 46].

Using UV-carcinogenic protocols and hairless mice (Skh-Hr-1), it was found that canthaxanthin, but not β-carotene or phytoene, could slow the development of subsequent tumors when administered immediately after the development of an initial tumor, i.e. a promotion stage effect [47]. However, phytoene was photoprotective when injected intraperitoneally for 10 weeks prior to exposure to a single, large, tumor-initiating fluence of UV. Further studies confirmed that oral (dietary) administration of 0.1% (w/w) of either β-carotene or canthaxanthin for 6 weeks prior to a single, tumor-initiating UV exposure and then continuing carotenoid administration during the progression phase, could significantly prevent the development of skin tumors [48]. When β-carotene was administered only before UV exposure, no significant protection was conferred. In this study, photoprotection was observed only when the pigment was administered during the progression phase of tumor development. In a rather complicated protocol in which various levels of dietary carotenoids were administered (complicated from the standpoint in that animals received the different levels of carotenoids in different 'run-in' feeding periods ranging from 4 days to 1 month, and in which the animal ages at time of irradiation ranged from 8 to 20 weeks), these investigators found that both canthaxanthin and β-carotene conferred significant protection against tumor development when the pigment was provided throughout the course of the study at a level of 0.07% (w/w) of the diet [16]. Lower levels did not exhibit protective properties. Using C3H/HeN mice in an UV protocol, a diet supplemented with 1% (w/w) canthaxanthin significantly reduced the average skin tumor burden per mouse [49]. Gensler et al. [50] also reported that feeding a diet supplemented with 1% canthaxanthin resulted in a significant reduction in the average skin tumor burden in UV-irradiated C3H/HeN mice, although the tumor latent period was unaffected.

At this point there seemed to be an overwhelming body of experimental evidence to substantiate the anti-UV-carcinogenic potential of carotenoids, especially β-carotene, that had arisen from the epidemiological studies. However, more recent studies employing C3H/HeN mice and a 1% β-carotene supplement failed to find a photoprotective effect of the carotenoid and, on close inspection of the data, an early exacerbation of carcinogenesis, with regard to tumor burden, occurred [51]. Further, when the influence of dietary supplementation of Skh-Hr-1 hairless mice with 0.07% (w/w) β-carotene, astaxanthin and lycopene was examined, β-carotene and astaxanthin significantly *exacerbated* UV-carcinogenic expression, with regard to both a shortened tumor latent period and increased tumor multiplicity [20]. Lycopene supplementation had no statistically significant effect on tumor expression.

From the foregoing, it is obvious that data from epidemiological, intervention and experimental studies concerning the influence of carotenoids (with focus upon β-carotene as this is the only carotenoid that has been employed in intervention studies) on carcinogenesis are in conflict. In the next section, an attempt is made to address issues that could provide insight into these disparate results.

Conclusion

Prior to the β-carotene intervention trials there was overwhelming epidemiological evidence that β-carotene would exhibit cancer-preventive properties. The strongest and most consistent evidence was reflected in an inverse association with lung cancer incidence and the consumption of foodstuffs rich in carotenoids, often reflected as blood levels of β-carotene, a marker of the former. Following the intervention trials in which β-carotene supplementation resulted in a significant increase in occurrence of lung cancer, or failed to provide protection against the subsequent occurrence of nonmelanoma skin cancer, several studies have attempted to address reasons for these unexpected results. One [52], a re-analysis of a case-control study, used an updated nutrient database from those used previously. Based upon consumption frequencies of 44 food items for which the database provided levels of individual carotenoids, these investigators found that low intake of both yellow-orange and dark-green vegetables was more predictive of elevated lung cancer than low intake of either β- or α-carotene. These authors suggested that supplemental β-carotene may have interfered with the utilization of other carotene isomers, carotenoids or phytochemicals. In this context, it is interesting to note that in experimental studies of chemically induced lung carcinogenesis, α-carotene administration resulted in a highly significant reduction in numbers of lung tumors per mouse whereas β-carotene had no significant effect [53]. A similar response was observed for chemically induced two-stage skin carcinogenesis.

Greenberg et al. [38] could find no alternative explanations that would entirely eliminate β-carotene as having a benefit in lowering the risk to overall mortality in the skin cancer prevention study but suggested that the decrease in mortality risk associated with higher initial plasma β-carotene levels might actually be a manifestation of the conjoint beneficial effects of a number of factors related to diet. They suggested that programs to encourage changes in dietary patterns that include greater consumption of fruit, vegetables and grains might prove more beneficially effective than altering the intake of individual micronutrients such as β-carotene.

In toto, these studies point to the precarious nature of epidemiological investigations that single out and promote anticancer effects for individual constituents of complex foodstuffs, and the words of Greenberg and Sporn [54] warrant reiteration: 'The disappointing results of the clinical trials of beta-carotene reaffirm the importance of solid scientific knowledge as the basis of disease prevention strategies'.

In this regard, the disparities between experimental UV carcinogenesis studies require resolution. In those cases where exacerbation of UV carcinogenesis by β-carotene has been observed, when taken with results from the clinical trials in which carotenoid supplementation has resulted in increased incidence of lung cancer, questions are raised that are more urgent than simply trying to find reasons for β-carotene's failure to convey protective effects. Does β-carotene supplementation pose a serious risk in fact and, if so, how?

Examination of experimental variables, from studies in which photoprotective effects of β-carotene on carcinogenesis were observed and studies in which exacerbation resulted, suggested that animal age and diet were the most probable causes of differences in response. It had previously been shown that antioxidant protection against UV carcinogenesis decreased with increasing animal age and that this effect might be related to the level of dietary antioxidant intake [55]. In the study in which exacerbation of UV carcinogenesis occurred in β-carotene-supplemented animals, aged animals were employed inasmuch as their responses to dietary supplementation with β-carotene might be more comparable to those in human adults, particularly when subjected to excessive oxidative stress from carcinogenic insult. Indeed, β-carotene supplementation resulted in significantly greater exacerbation of UV carcinogenesis in aged than in young animals [56]. However, the difference in animal age was not completely responsible for the different photocarcinogenic responses elicited by β-carotene supplementation, as young animals exhibited a significant exacerbation with respect to tumor multiplicity.

The only obvious remaining variable between studies in which β-carotene was shown to protect and those in which it exacerbated UV carcinogenesis was diet. Earlier studies in which a photoprotective effect was observed employed commercial, closed-formula rations whereas those studies in which β-carotene failed to exhibit photoprotection, or resulted in exacerbation, employed semidefined diets. In a carefully controlled experiment comparing the effects of β-carotene-supplemented closed-formula and semidefined diets, no significant influence on tumor latent period or tumor multiplicity was observed with the closed-formula ration. However, the β-carotene-supplemented semidefined diet exerted a statistically significant exacerbation of UV carcinogenesis with respect to both parameters [57]. These data suggest that response to carotenoid supplementation might well depend on the presence

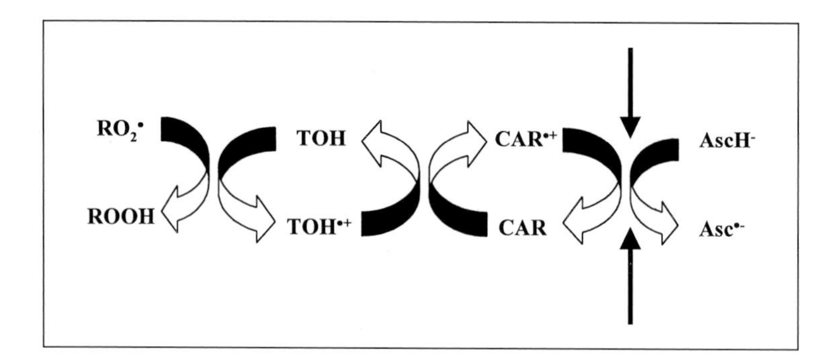

Fig. 4. Proposed carotenoid radical cation repair mechanism by ascorbic acid. Either an ascorbic acid deficiency, as in the case of smokers (top arrow), or as a consequence of an increased level of ascorbic acid radical resulting from oxidative stress upon carcinogen exposure (lower arrow), would lead to an increased level of the unrepaired carotenoid radical cation. This could result in a 'prooxidant-like' state. The unrepaired carotenoid radical cation could damage membranes by direct action or indirectly by blocking the effective antioxidant activity of α-tocopherol.

and interaction with other dietary factors, factors that were either absent or present in ineffectual concentrations in semidefined diets. In this regard, the schema in figure 2 suggests such an interaction. Smokers, even nonsmokers exposed to passive smoke, exhibit markedly reduced plasma vitamin C levels that reflect severe oxidant exposure [58]. In fact, smokers also exhibit a lower β-carotene load that is attributed to the large number of free radicals in cigarette smoke (estimated to be around 10^{15} per puff) [59]. Thus, conditions of excessive oxidative stress, in the case of potential carcinogen exposure, and diminished ascorbic acid levels, particularly in aged animals or adult humans, could lead to an ineffective carotenoid radical repair mechanism and eventually to the equivalent of a prooxidant state. Mice have no known vitamin C requirement for normal growth and development [60]. The semidefined diet used in those studies in which exacerbation occurred contained vitamin C whereas the closed-formula rations, in which photoprotective effects were observed, did not. Either condition, i.e. lack of vitamin C in smokers or additional vitamin C supplied under oxidative stress of UV irradiation (in which case vitamin C may act as a prooxidant), would be expected to inhibit carotenoid radical cation repair and stoichiometrically alter the repair mechanism to one of prooxidative influence (fig. 4). Alternatively, other carotenoids or phytochemicals present in closed-formula diets might conjointly act to determine the UV-carcinogenic response in the presence of β-carotene.

Regardless of the mechanism, there is a growing body of evidence that diet can have a pronounced influence on β-carotene-modulated carcinogenesis and underscores the potential risk of β-carotene supplementation. Such a concern is reflected in the conclusions of the IARC working group [1]: 'Until further insight is gained, β-carotene should not be recommended for use in cancer prevention in the general population and it should not be assumed that β-carotene is responsible for the cancer protecting effects of diets rich in carotenoid-containing fruits and vegetables.' It was the consumption of such foodstuffs on which the anticancer effects of carotenoids were initially based [11]. In this context, it may be prudent to weigh the potential risk-benefit when employing β-carotene therapeutically for other disorders as well.

Acknowledgements

Work of H.S.B. has been supported, in part, by a grant from the American Institute for Cancer Research.

References

1 IARC Working Group on the Evaluation of Cancer-Preventive Agents: IARC Handbooks of Cancer Prevention: Carotenoids. Lyon, IARC, 1998, vol 2.

2 Sistrom WR, Griffiths M, Stanier RY: A note on the porphyrins excreted by the blue-green mutant of *Rhodopseudomonas spheroides*. J. Cell Comp Physiol 1956;48:459–515.

3 Mathews-Roth MM, Sistrom WR: Function of carotenoid pigments in non-photosynthetic bacteria. Nature 1959;184:1892–1893.

4 Kesten BM: Urticaria solare. Arch Dermatol Syphilol 1951;64:221–223.

5 Mathews-Roth MM: Carotenoids in medical applications; in Bauernfeind JC (ed): Carotenoids as Colorants and Vitamin A Precursors. New York, Academic Press, 1981, pp 755–785.

6 Mathews-Roth MM: Beta-carotene therapy for erythropoietic protoporphyria and other photosensitivity diseases; in Regan JD, Parrish JA (eds): The Science of Photobiology. New York, Plenum Press, 1982, pp 409–440.

7 Mathews MM: Protective effect of β-carotene against lethal photosensitization by haematoporphyrin. Nature 1964;203:1092.

8 Mathews-Roth MM, Pathak MA, Fitzpatrick TB, Haber LH, Kass EH: Beta-carotene therapy for erythropoietic protoporphyria and other photosensitivity diseases. Arch Dermatol 1977;113: 1229–1232.

9 Klaui H, Bauernfeind JC: Carotenoids as food colors; in Bauernfeind JC (ed): Carotenoids as Colorants and Vitamin A Precursors. New York, Academic Press, 1981, pp 47–317.

10 Shekelle RB, Liu S, Raynor WJ, Lepper M, Maliza C, Rossof AH, Pau O, Shryock AM, Stamler J: Dietary vitamin A and risk of cancer in the Western Electric study. Lancet 1991;ii: 1185–1190.

11 Peto R, Doll R, Sporn MB: Can dietary β-carotene materially reduce human cancer rates? Nature 1981;290:201–208.

12 Singh VN: Role of β-carotene in disease prevention with special reference to cancer; in Ong ASH, Packer L (eds): Lipid-Soluble Antioxidants: Biochemistry and Clinical Applications. Basel, Birkhäuser, 1992, pp 208–227.

13 Kune GA, Bannerman S, Field B, Watson LF, Cleland H, Merenstein D, Vitetta L: Diet, alcohol, smoking, serum β-carotene, and vitamin A in male nonmelanocytic skin cancer patients and controls. Nutr Cancer 1992;18:237–244.

14 Epstein JH: Effects of β-carotene on UV-induced skin cancer formation in the hairless mouse skin. Photochem Photobiol 1977;25:211–213.

15 Mathews-Roth MM: Antitumor activity of β-carotene, canthaxanthin, and phytoene. Oncology 1982;38:33–37.

16 Mathews-Roth MM, Krinsky NI: Carotenoid dose level and protection against UV-B-induced skin tumors. Photochem Photobiol 1985;42:35–38.

17 Mathews-Roth MM: β-Carotene: Clinical aspects; in Spiller GA, Scala J (eds): New Protective Role for Selected Nutrients. New York, Liss, 1989, pp 17–38.

18 Greenberg ER, Baron JA, Stukel TA, Stevens MM, Mandel JS, Spencer SK, Elias PM, Lowe N, Nierenberg DW, Bayrd G, Vance JC, Freeman DH, Clendenning WE, Kwan T, and the Skin Cancer Prevention Study Group: A clinical trial of β-carotene to prevent basal-cell and squamous-cell cancers of the skin. N Engl J Med 1990;323:789–795.

19 The α-Tocopherol, β-Carotene Cancer Prevention Study Group: The effect of vitamin E and β-carotene on the incidence of lung cancer and other cancers in male smokers. N Engl J Med 1996; 330:1029–1035.

20 Black HS: Radical interception by carotenoids and effects on UV carcinogenesis. Nutr Cancer 1998; 31:212–217.

21 Foote CS, Denny RW: Chemistry of singlet oxygen. VII. Quenching by β-carotene. J Am Chem Soc 1968;90:6233–6235.

22 Foote CS, Chang YC, Denny RW: Chemistry of singlet oxygen. XI. Cis-trans isomerization of carotenoids by singlet oxygen and a probable quenching mechanism. J Am Chem Soc 1970;92: 5218–5219.

23 Foote CS, Chang YC, Denny RW: Chemistry of singlet oxygen. X. Carotenoid quenching parallels biological properties. J Am Chem Soc 1970;92:5216–5218.

24 Mathews-Roth MM, Wilson T, Fujimori E, Krinsky NI: Carotenoid chromophore length and protection against photosensitization. Photochem Photobiol 1974;19:217–222.

25 Krinsky NI, Deneke SM: Interaction of oxygen and oxy-radicals with carotenoids. J Natl Cancer Institute 1982;69:205–210.

26 Palozza P, Moualla S, Krinsky NI: Effects of β-carotene and α-tocopherol on radical initiated peroxidation of microsomes. Free Radic Biol Med 1992;13:127–136.

27 Burton GW, Ingold KU: Beta-carotene: An unusual type of lipid antioxidant. Science 1984;224: 569–573.

28 Hill TJ, Land EJ, McGarvey DJ, Schalch W, Tinkler JH, Truscott TG: Interactions between carotenoids and the $CCL_3O_2^{\cdot}$ radical. J Am Chem Soc 1995;117:8322–8326.

29 Tinkler JH, Tavender SM, Parker AW, McGarvey DJ, Mulroy L, Truscott TG: Investigation of carotenoid radicals and triplet states by laser flash photolysis and time-resolved resonance Raman spectroscopy: Observation of competitive energy and electron transfer. J Am Chem Soc 1996;118: 1756–1761.

30 Dawe EA, Land EJ: Radical ions derived from photosynthetic polyenes. J Chem Soc (Faraday 1) 1975;71:2162–2169.

31 Anderson HJ, Chen H, Pellett LJ, Tappel AL: Ferrous-iron-induced oxidation in chicken liver slices as measured by hemichrome formation and thiobarbituric acid-reactive substances: Effects of dietary vitamin E and β-carotene. Free Radic Biol Med 1993;15:37–48.

32 Truscott TG: β-Carotene and disease: A suggested pro-oxidant and anti-oxidant mechanism and speculations concerning its role in cigarette smoking. J Photochem Photobiol B Biol 1996;35: 233–235.

33 Edge R, Land EJ, McGarvey D, Mulroy L, Truscott TG: Relative one-electron reduction potentials of carotenoid radical cations and the interactions of carotenoids with the vitamin E radical cation. J Am Chem Soc 1998;120:4087–4090.

34 Bohm F, Edge R, Land EJ, McGarvey DJ, Truscott TG: Carotenoids enhance vitamin E antioxidant efficiency. J Am Chem Soc 1997;119:621–622.

35 Packer JE, Slater TF, Wilson RL: Direct observation of a free radical interaction between vitamin E and vitamin C. Nature 1979;278:737–738.
36 Valgimigli L, Lucarini M, Pedulli GF, Ingold KU: Does β-carotene really protect vitamin E from oxidation? J Am Chem Soc 1998;119:8095–8096.
37 Pryor WA, Cornicelli JA, Devall LJ, Tait B, Trivedi BK, Witiak DT, Wu M: A rapid screening test to determine the antioxidant potencies of natural and synthetic antioxidants. J Org Chem 1993;58: 3521–3532.
38 Greenberg ER, Baron JA, Karagas MR, Strukel TA, Nierenberg DW, Stevens MM, Mandel JS, Haile RW: Mortality associated with low plasma concentration of beta carotene and the effect of oral supplementation. JAMA 1996;275:699–703.
39 Karagas MR, Greenberg ER, Nierenberg D, Stukel TA, Morris JS, Stevens MM, Baron JA: Risk of squamous cell carcinoma of the skin in relation to plasma selenium, α-tocopherol, β-carotene, and retinol: A nested case-control study. Cancer Epidemiol Biomarkers Prev 1997;6:25–29.
40 Stryker WS, Stampfer MJ, Stein EA, Kaplan L, Louis TA, Sober A, Willett WC: Diet, plasma levels of β-carotene and alpha-tocopherol, and risk of malignant melanoma. Am J Epidemiol 1990; 131:597–611.
41 Kirkpatrick CS, White E, Lee JAH: Case-control study of malignant melanoma in Washington state. Am J Epidemiol 1994;139:869–880.
42 Breslow RA, Alberg AJ, Helzlsouer KJ, Bush TL, Norkus EP, Morris JS, Spate VE, Comstock GW: Serological precursors of cancer: Malignant melanoma, basal and squamous cell skin cancer, and prediagnostic levels of retinol, β-carotene, lycopene, α-tocopherol, and selenium. Cancer Epidemiol Biomarkers Prev 1995;4:837–842.
43 Bayerl C, Schwarz B, Jung EG: Beneficial effect of beta-carotene in dysplastic nevus syndrome – A randomized trial (abstract 068). 8th Congr Eur Soc Photobiol, Granada, 1999, p 89.
44 Mathews-Roth MM, Pathak MA, Parrish J, Fitzpatrick TB, Kass EH, Toda K, Clemens W: A clinical trial of the effects of oral beta-carotene on the responses of human skin to solar radiation. J Invest Dermatol 1972;59:349–353.
45 Mathews-Roth MM, Pathak MA: Phytoene as a protective agent against sunburn (>280 nm) radiation in guinea pigs. Photochem Photobiol 1975;21:261–263.
46 Santamaria L, Bianchi A, Arnaboldi A, Andreoni L: Prevention of the benzo(α)-pyrene photocarcinogenic effect by β-carotene and canthaxanthin. Med Biol Environ 1981;9:113–120.
47 Mathews-Roth MM: Carotenoid pigment administration and delay in development of UV-B-induced tumors. Photochem Photobiol 1983;37:509–511.
48 Mathews-Roth MM, Krinsky NI: Carotenoids affect development of UV-B induced skin cancer. Photochem Photobiol 1987;46:507–509.
49 Rybski JA, Grogan TM, Aickin M, Gensler HL: Reduction of murine cutaneous UVB-induced tumor-infiltrating T lymphocytes by dietary canthaxanthin. J Invest Dermatol 1991;97:892–897.
50 Gensler HL, Aickin M, Peng YM: Cumulative reduction of primary skin tumor growth in UV-irradiated mice by the combination of retinyl palmitate and canthaxanthin. Cancer Lett 1990;53: 27–31.
51 Gensler HL, Magdaleno M: Topical vitamin E inhibition of immunosuppression and tumorigenesis induced by ultraviolet irradiation. Nutr Cancer 1991;15:97–106.
52 Ziegler RG, Colavito EA, Hartge P, McAdams MJ, Schoenberg JB, Mason TJ, Fraumeni JF Jr: Importance of α-carotene, β-carotene, and other phytochemicals in the etiology of lung cancer. J Natl Cancer Inst 1996;88:612–615.
53 Nishino H, Murakoshi M, Kitana H, Iwasaki R, Tanaka Y, Tsushina M, Matsuno T, Okabe H, Okuzumi J, Hasegawa T, Takayasu J, Satomi Y, Tokuda H, Nishino A, Iwashima A: Anti-tumor and anti-tumor promoting activity of α- and β-carotene; in Ong ASH, Packer L (eds): Lipid-Soluble Antioxidants: Biochemistry and Clinical Applications. Basel, Birkhäuser, 1992, pp 228–242.
54 Greenberg ER, Sporn MB: Antioxidant vitamins, cancer, and cardiovascular disease. N Engl J Med 1996;334:1189–1190.
55 Black HS, McCann V, Thornby JI: Influence of animal age upon antioxidant-modified UV carcinogenesis. Photobiochem Photobiophys 1982;4:107–118.

56 Black HS, Okotie-Eboh G, Gerguis J: Aspects of β-carotene-potentiated photocarcinogenesis. Photochem Photobiol 1999;69:32S.
57 Black HS, Okotie-Eboh G, Gerguis J: Diet potentiates the UV-carcinogenic response to β-carotene. Nutr Cancer, accepted for publication.
58 Tribble DL, Giuliana LJ, Fortmann SP: Reduced plasma ascorbic acid concentrations in nonsmokers regularly exposed to environmental tobacco smoke. Am J Clin Nutr 1993;58:886–890.
59 Handelman GJ, Packer L, Cross CE: Destruction of tocopherols, carotenoids, and retinol in human plasma by cigarette smoke. Am J Clin Nutr 1996;63:559–565.
60 Nutrient Requirements of Laboratory Animals, ed 3. Washington, National Research Council/ National Academy of Sciences, 1978, pp 38–53.

Dr. Homer S. Black, Photobiology Laboratory, Research Service,
Veterans Affairs Medical Center, 2002 Holcombe Blvd., Houston, TX 77030 (USA)
Tel. +1 713 794 7637, Fax +1 713 794 7938, E-Mail hblack@bcm.tmc.edu

Thiele J, Elsner P (eds): Oxidants and Antioxidants in Cutaneous Biology.
Curr Probl Dermatol. Basel, Karger, 2001, vol 29, pp 157–164

Protective Effects of Topical Antioxidants in Humans

Frank Dreher, Howard Maibach

Department of Dermatology, University of California, San Francisco, Calif., USA

As the outermost barrier of the body, the skin is directly exposed to a prooxidative environment, including ultraviolet radiation (UVR) and air pollutants [1]. These external inducers of oxidative attack lead to the generation of reactive oxygen species (ROS) and other free radicals. To counteract the harmful effects of ROS, the skin is equipped with antioxidant systems consisting of a variety of primary (preventive, e.g. vitamin C) and secondary (interceptive, e.g. vitamin E) antioxidants forming an 'antioxidant network'. Antioxidants intervene at different levels of oxidative processes, e.g. by scavenging free radicals and lipid peroxyl radicals, by binding metal ions or by removing oxidatively damaged biomolecules [2].

The antioxidant network is responsible for maintaining an equilibrium between pro- and antioxidants. However, the antioxidant defense in cutaneous tissues can be overwhelmed by an increased exposure to exogenous sources of ROS. Such a disturbance of the prooxidant/antioxidant balance may result in oxidative damage of lipids, proteins and DNA, and has been termed 'oxidative stress' [3].

The important role of ROS in UVR-induced skin damage is well documented. UVR-induced skin damage includes acute reactions, such as erythema, edema, followed by exfoliation, tanning and epidermal thickening. Premature skin aging ('photoaging') and carcinogenesis are believed to be consequences of chronic UVR exposure. A further environmental stressor is tropospheric ozone (O_3). While ozone in the upper atmosphere (stratosphere) occurs naturally and protects skin by filtering out harmful solar UVR, O_3 at ground level (troposphere) is a noxious, highly reactive air pollutant. Recently, a series of studies were published investigating the impact of O_3 on skin antioxidants.

Since O_3 levels are frequently highest in areas where exposure to UVR is also high, the co-exposure to O_3 and UVR in photochemical smog could result in additive if not even synergistic harmful effects on the skin.

This review focuses on the currently available knowledge on the protective effects of topical antioxidants in humans against exogenous oxidative stressors.

Effect of Environmental Stressors on the Constitutive Skin Antioxidants

Human skin contains the lipophilic antioxidants vitamin E (tocopherols and tocotrienols), ubiquinones (coenzyme Q) and carotenoids, as well as the hydrophilic antioxidants vitamin C (ascorbate), uric acid (urate) and glutathione (GSH) [1]. Generally, higher antioxidant concentrations were found in the epidermis as compared to the dermis. Some antioxidants are also present in the stratum corneum. α-Tocopherol was shown to be the predominant antioxidant in human stratum corneum, whereas the outermost layers seem to contain lower amounts than the layers in closer proximity to the nucleated epidermis [4].

Particularly, the antioxidants contained in the stratum corneum were demonstrated to be susceptible to UVR. For example, a single suberythemal dose of UVR depleted human stratum corneum α-tocopherol by almost half, while dermal and epidermal α-tocopherol were only depleted at much higher doses [4]. The high susceptibility of stratum corneum vitamin E to UVR may be, at least in part, due to a lack of co-antioxidants in this outermost skin layer. Ubiquinol 10, the most abundant ubiquinone found in human skin, was undetectable in human stratum corneum. Additionally, ascorbate, the major hydrophilic co-antioxidant that is capable of recycling photooxidized α-tocopherol [5], seems to be present only at very low levels in human stratum corneum. Consequently, direct depletion of α-tocopherol and formation of its radical may also affect other endogenous antioxidant pools. In addition to direct depletion by UVR, skin α-tocopherol levels may be consumed as a consequence of its chain-breaking antioxidant action.

Also the hydrophilic antioxidants were shown to be sensitive to UVR. However, ascorbic and uric acid were less susceptible to UVR than α-tocopherol or ubiquinol 10 as was shown using cultured human skin equivalents [6]. Further, epidermal GSH levels in hairless mice were depleted by 40% within minutes after UVB exposure but returned to normal levels after half an hour [7]. Moreover, skin contains enzymatic antioxidants such as catalase, superoxide dismutase, GSH peroxidase and GSSG reductase, which were also shown to be susceptible to UVR [1].

Recently, the effects of the air pollutant ozone on skin antioxidants have been reported [8, 9]. Similarly, as found for UVR exposure, the stratum corneum is the most susceptible skin layer for ozone-induced depletion of lipophilic and hydrophilic antioxidants as was demonstrated using hairless mice. It seems that ozone itself is too reactive to penetrate deeply into the skin and reacts therefore predominantly with the skin barrier lipids and proteins in the outermost epidermis.

Role of Antioxidants in the Photoprotection of Human Skin

Topical Application of a Single Antioxidant

Apart from using chemical and/or physical sunscreens to diminish the intensity of UVR reaching the skin, supplementation of the skin with antioxidants and thereby strengthening its antioxidative capacity is an emerging approach in limiting ROS-induced skin damage [1]. Topical application of antioxidants, such as vitamin E, provides an efficient means of increasing antioxidant tissue levels in human epidermis [10]. As the most susceptible skin layer for UVR- and ozone-induced depletion of cutaneous antioxidants, the stratum corneum may particularly benefit from an increased antioxidant capacity due to topical supplementation.

Vitamin E

The photoprotective effects of vitamin E (α-tocopherol) have been studied extensively [1]. However, most studies were performed on animals, and only few studies exist investigating the photoprotective effects of topically applied vitamin E in humans. Significantly reduced acute skin responses such as erythema and edema, sunburn cell formation, lipid peroxidation, DNA adduct formation, immunosuppression as well as UVA-induced binding of photosensitizers were demonstrated when vitamin E was applied before UVR exposure. As shown in animal studies, skin wrinkling and skin tumor incidence due to chronic UVR exposure seem also to be diminished by topical vitamin E.

A human study proved that an alcoholic lotion containing 2% α-tocopherol significantly diminished the erythemal responses as assessed by measuring the chromameter a* value (representing redness of the erythema) and the dermal blood flow [11]. The lotion was applied 30 min before UVR exposure at a dose of 2 mg/cm^2. The lotion had no significant sunscreening properties as an in vitro determination of the sun protection factor demonstrated. Therefore, the observed photoprotective effect of this lotion may be attributed to the antioxidant properties of α-tocopherol. The fate of topically applied α-tocopherol in UVB-irradiated skin is, however, still debated, since investigations on the UVB-induced photooxidation of α-tocopherol in liposomes indicated that α-tocopherol might also act as a sunscreen [12]. Different vitamin E doses/concentrations and differing experimental setups used might explain the nonconforming results regarding the determination of vitamin E's sunscreening properties.

Vitamin E esters, particularly vitamin E acetate, succinate and linoleate, were also shown to be promising agents in reducing UVR-induced skin damage [1]. However, their photoprotective effects appear to be less pronounced as compared to vitamin E. Vitamin E esters need to be hydrolyzed during skin absorption to show antioxidant activity, but it seems that the bioconversion of vitamin E acetate to its active antioxidative form α-tocopherol is slow and occurs only to a minor extent. As was demonstrated during a human study, twice daily application of an α-tocopherol-acetate-containing cream over 3 months did not result in any evidence of conversion within skin to its free form although it was substantially absorbed [13]. Hence, the less pronounced photoprotective effects of topically applied vitamin E acetate after a single application might be explained by a limited bioavailability of the ester-cleaved form during oxidative stress in the superficial skin layers.

Some evidence exists, however, that the bioconversion of vitamin E acetate into vitamin E might be enhanced due to UVR exposure [14]. UVB exposure was demonstrated to cause an increase in esterase activity in the epidermis. The study was performed by applying deuterated α-tocopherol acetate on hairless mice in order to distinguish between endogenous and deuterium-labeled α-tocopherol resulting from hydrolysis.

Vitamin C

Few studies described the photoprotective effects of vitamin C after topical application. Using a porcine skin model, it was demonstrated that topically applied vitamin C is only effective when formulated at high concentration in an appropriate vehicle [15]. On the other hand, as could be shown in a human study, 5% vitamin C incorporated into an alcoholic lotion was unable to induce any significant photoprotective effect when applied at a dose of 2 mg/cm^2 30 min before UVR irradiation [11]. Vitamin C does not act as sunscreen.

The modest photoprotective effect of topically applied vitamin C may be explained by its instability and ease of oxidation in aqueous vehicles [16]. Vitamin C might be protected from degradation by selecting appropriate, sophisticated vehicles. Furthermore, lipophilic and stable vitamin C derivatives, such as its palmityl, succinyl or phosphoryl esters, might be promising compounds providing increased photoprotection as compared to vitamin C [17].

As described for vitamin E esters, most of these compounds must be hydrolyzed to vitamin C to be effective as antioxidants.

Other Antioxidants

Apart from vitamins E and C, and derivatives thereof, diverse other topically applied compounds with antioxidative properties were shown to efficiently diminish photodamage [1]. Particularly flavonoids were reported to reduce acute and chronic skin damage after UVR exposure. Flavonoids (e.g. apigenin, catechin, epicatechin, α-glycosylrutin and silymarin) are polyphenolic compounds which can be extracted from plants. A tropical fern extract reduced erythema as well as UVA-induced immediate pigment darkening and delayed tanning when applied to human skin before UVR exposure [18].

Thiols, such as N-acetylcysteine and derivatives, are another important group of potent radical scavengers, and their topical administration was shown to diminish UVA-induced binding of photosensitizers to epidermal lipids and DNA in rats [1]. In addition, α-lipoic acid may also be an interesting antioxidant to protect against oxidative stressors when applied onto skin [19].

The pineal hormone melatonin (N-acetyl-5-methoxytryptamine) is assumed to be a further potent antioxidant. Its beneficial effect in reducing UVR-induced erythema could be established during human studies [11, 20, 21]. Apart from melatonin's antioxidant properties, its dose-dependent sunscreening properties, as well as its supposed immunomodulatory function might have contributed to the observed photoprotective effects.

Topical Application of Antioxidant Combinations

The cutaneous antioxidant system is complex and far from being completely understood. As mentioned above, the system is interlinked and operates as an antioxidant network [1]. α-Tocopherol is readily regenerated from its radical at the expense of reductants like ascorbate. Ascorbate itself can be regenerated by GSH. Thus, an enhanced photoprotective effect may be obtained by applying appropriate combinations of antioxidants.

As was shown in a human study, the co-application of vitamin E and vitamin C resulted in a much more pronounced photoprotective effect as compared to the application of one single antioxidant alone in the identical vehicle [11]. The antioxidant mixture was applied half an hour before UVR irradiation at a dose of 2 mg/cm^2. The vehicle was selected in order to dissolve 2% vitamin E, 5% vitamin C, and contained a nonionic surfactant. The determination of the in vitro sun protection factor of this mixture resulted in no sunscreening properties. Consequently, the photoprotective effect of such a mixture may be based on an antioxidant mechanism. And, as was demonstrated in the same study, the most dramatic improvement resulted from the co-formulation of melatonin together with vitamin E and vitamin C. It may be speculated that synergistic interactions between melatonin and the vitamins E and C could have contributed to the observed, significantly increased photoprotective effects.

Distinct mixtures of topically applied antioxidants were also shown to be helpful in reducing the development and the severity of experimentally induced polymorphous light eruption in humans [22]. A combination consisting of α-glycosylrutin, ferulic acid and tocopheryl acetate incorporated into a conventional oil-in-water emulsion was applied twice daily for 1 week prior to the photoprovocation with UVA on 4 consecutive days. The authors showed that the sunscreening effect of the antioxidant-containing emulsion was negligible and that the photoprotective effect observed was due to reduction of UVA-induced oxidative stress.

Furthermore, 5% vitamin E linoleate combined with 1% magnesium ascorbyl phosphate incorporated into an oil-in-water emulsion, containing furthermore 0.03% butylated hydroxytoluene as well as 0.01% nordihydroguaradinic acid (also representing synthetic antioxidants), significantly reduced UVR-induced erythema in humans [23].

Topical Application of Antioxidants after UVR Exposure

Whereas the photoprotective effect of topical antioxidants applied before UVR exposure has been recognized, the effect of these compounds administered after irradiation is less obvious. Diminished erythema formation was reported when antioxidants were topically administered after UVR exposure in humans [21, 24]. However, these findings are in contrast to other human studies which were not able to demonstrate any diminished UVR-related skin damage when antioxidants were applied after irradiation [20, 25]. As was shown, neither vitamin E nor vitamin C nor melatonin nor combinations thereof led to a significantly lowered erythema formation when administered after UVR exposure [25]. It seems, therefore, that UVR-induced ROS formation and the subsequent reaction of ROS with skin biomolecules leading consequently to acute skin damage is a very rapid process. Hence, since antioxidants applied after irradiation possibly do not reach the site of action (e.g. superficial skin layers) in relevant amounts during the occurrence of oxidative stress, they do not significantly reduce UVR-induced erythema formation as compared to their vehicles. And, such skin damage may be more efficiently treated on an inflammatory level by classical anti-inflammatory drugs.

Summary and Conclusion

Human studies have convincingly demonstrated pronounced photoprotective effects of 'natural' and synthetic antioxidants when applied topically before UVR exposure. Particularly with respect to UVB-induced skin damage such as erythema formation, the photoprotective effects of antioxidants are significant when applied in distinct mixtures in appropriate vehicles. Topical application of such combinations may result in a sustained antioxidant capacity of the skin, possibly due to antioxidant synergisms. And, since UVA-induced skin alterations are believed to be largely determined by oxidative processes [26], topical administration of antioxidants might be particularly promising [27, 28]. In fact, topical application of antioxidants or antioxidant mixtures resulted in a remarkable increase in the minimal dose to induce immediate pigment darkening after UVA exposure [18, 23] and diminished the severity of UVA-induced photodermatoses [22] in humans.

In conclusion, regular application of skin care products containing antioxidants may be of the utmost benefit in efficiently preparing our skin against exogenous oxidative stressors occurring during daily life. Furthermore, sunscreening agents may also benefit from combination with antioxidants resulting in increased safety and efficacy of such photoprotective products [11, 29].

References

1 Thiele JJ, Dreher F, Packer L: Antioxidant defense systems in skin; in Elsner P, Maibach H, Rougier A (eds): Drugs vs Cosmetics: Cosmeceuticals? New York, Dekker, 2000, pp 145–187.

2 Briviba K, Sies H: Nonenzymatic antioxidant defense systems; in Frei B (ed): Natural Antioxidants in Human Health and Disease. New York, Academic Press, 1994.

3 Sies H: Introductory remarks; in Sies H (ed): Oxidative Stress. Orlando, Academic Press, 1985, pp 1–7.

4 Thiele JJ, Traber MG, Packer L: Depletion of human stratum corneum vitamin E: An early and sensitive in vivo marker of UV-induced photooxidation. J Invest Dermatol 1998;110:756–761.

5 Kagan V, Witt E, Goldman R, Scita G, Packer L: Ultraviolet light-induced generation of vitamin E radicals and their recycling: A possible photosensitizing effect of vitamin E in skin. Free Radic Res Commun 1992;16:51–64.

6 Podda M, Traber MG, Weber C, Yan LJ, Packer L: UV-irradiation depletes antioxidants and causes oxidative damage in a model of human skin. Free Radic Biol Med 1998;24:55–65.

7 Connor MJ, Wheeler LA: Depletion of cutaneous glutathione by ultraviolet radiation. Photochem Photobiol 1987;47:239–245.

8 Thiele JJ, Podda M, Packer L: Tropospheric ozone: An emerging environmental stress to skin. Biol Chem 1997;378:1299–1305.

9 Weber SU, Thiele JJ, Cross CE, Packer L: Vitamin C, uric acid, and glutathione gradients in murine stratum corneum and their susceptibility to ozone exposure. J Invest Dermatol 1999;113:1128–1132.

10 Wester RC, Maibach HI: Absorption of tocopherol into and through human skin. Cosmet Toiletr 1997;112:53–57.

11 Dreher F, Gabard B, Schwindt DA, Maibach HI: Topical melatonin in combination with vitamins E and C protects skin from UV-induced erythema: A human study in vivo. Br J Dermatol 1998; 139:332–339.
12 Kramer KA, Liebler DC: UVB induced photooxidation of vitamin E. Chem Res Toxicol 1997;10: 219–224.
13 Alberts DS, Goldman R, Xu MJ, Dorr RT, Quinn J, Welch K, Guillen-Rodriguez J, Aickin M, Peng YM, Loescher L, Gensler H: Disposition and metabolism of topically administered α-tocopherol acetate: A common ingredient of commercially available sunscreens and cosmetics. Nutr Cancer 1996;26:193–201.
14 Kramer-Stickland KA, Liebler DC: Effect of UVB on hydrolysis of α-tocopherol acetate to α-tocopherol in mouse skin. J Invest Dermatol 1998;111:302–307.
15 Darr D, Combs S, Dunston S, Manning T, Pinnell S: Topical vitamin C protects porcine skin from ultraviolet radiation-induced damage. Br J Dermatol 1992;127:247–253.
16 Austria R, Semenzato A, Bettero A: Stability of vitamin C derivatives in solution and topical formulations. J Pharm Biomed Anal 1997;15:795–801.
17 Kobayashi S, Takehana M, Itoh S, Ogata E: Protective effect of magnesium-*L*-ascorbyl-2 phosphate against skin damage induced by UVB irradiation. Photochem Photobiol 1996;64:224–228.
18 González S, Pathak MA, Cuevas J, Villarrubia VG, Fitzpatrick TB: Topical or oral administration with an extract of *Polypodium leucotomos* prevents acute sunburn and psoralen-induced phototoxic reactions as well as depletion of Langerhans cells. Photodermatol Photoimmunol Photomed 1997; 13:50–60.
19 Fuchs J, Milbradt R: Antioxidant inhibition of skin inflammation induced by reactive oxidants: Evaluation of the redox couple dihydrolipoate-lipoate. Skin Pharmacol 1994;7:278–284.
20 Bangha E, Elsner P, Kistler GS: Suppression of UV-induced erythema by topical treatment with melatonin (N-acetyl-5-methoxytryptamine): Influence of the application time point. Dermatology 1997;195:248–252.
21 Bangha E, Elsner P, Kistler GS: Suppression of UV-induced erythema by topical treatment with melatonin (N-acetyl-5-methoxytryptamine): A dose response study. Arch Dermatol Res 1996;288: 522–526.
22 Hadshiew I, Stäb F, Untiedt S, Bohnsack K, Rippke F, Hölzle E: Effects of topically applied antioxidants in experimentally provoked polymorphous light eruption. Dermatology 1997;195: 362–368.
23 Muizzuddin N, Shakoori AR, Marenus KD: Effect of antioxidants and free radical scavengers on protection of human skin against UVB, UVA and IR irradiation. Skin Res Technol 1999;5:260–265.
24 Montenegro L, Bonina F, Rigano L, Giogilli S, Sirigu S: Protective effect evaluation of free radical scavengers on UVB induced human cutaneous erythema by skin reflectance spectrophotometry. Int J Cosmet Sci 1995;17:91–103.
25 Dreher F, Denig N, Gabard B, Schwindt DA, Maibach HI: Effect of topical antioxidants on UV-induced erythema formation when administered after exposure. Dermatology 1999;198:52–55.
26 Tyrrell RM: UVA (320–380 nm) radiation as an oxidative stress; in Sies H (ed): Oxidative Stress: Oxidants and Antioxidants. London, Academic Press, 1991, pp 57–83.
27 Evelson P, Ordóñez CP, Llesuy S, Boveris A: Oxidative stress and in vivo chemiluminescence in mouse skin exposed to UVA radiation. J Photochem Photobiol B Biol 1997;38:215–219.
28 Clement-Lacroix P, Michel L, Moysan A, Morlière P, Dubertret L: UVA-induced immune suppression in human skin: Protective effects of vitamin E in human epidermal cells in vitro. Br J Dermatol 1996;134:77–84.
29 Darr D, Dunston S, Faust H, Pinnell S: Effectiveness of antioxidants (vitamin C and E) with and without sunscreens as topical photoprotectants. Acta Dermatol Venereol 1996;76:264–268.

Prof. Howard Maibach, Department of Dermatology, University of California, School of Medicine, Box 0989, Surge 110, San Francisco, CA 94143 (USA)
Tel. +1 415 476 24 68, Fax +1 415 753 53 04, E-Mail himjlm@itsa.ucsf.edu

Thiele J, Elsner P (eds): Oxidants and Antioxidants in Cutaneous Biology.
Curr Probl Dermatol. Basel, Karger, 2001, vol 29, pp 165–174

The Antioxidative Potential of Melatonin in the Skin

Tobias W. Fischer, Peter Elsner

Department of Dermatology and Allergology, Friedrich Schiller University, Jena, Germany

Melatonin (N-acetyl-5-methoxytryptamine) is an indole hormone produced by the pineal gland and exhibits primarily biorhythmic functions such as the regulation of seasonal biorhythmic functions, reproduction, daily sleep induction in dependency of light perception, aging and modulation of immunobiological reactions [1]. This hormone is present in the human blood at concentrations of about 10 pg/ml during daytime and increases at night up to a peak of 250 pg/ml between 2.00 and 4.00 a.m. [2]. The hormonal functions are transmitted by the specific melatonin receptors Mel-1 and Mel-2 [3]. Melatonin is a highly lipophilic molecule which penetrates through cell membranes and reaches intracellular compartments (fig. 1). Because melatonin is present in almost all organisms from the primitive unicellular organism to the human, it was supposed that it has other functions in addition to the regulation of hormonal cycles [4]. Indeed, it was found to function as an antioxidant due to its strong radical scavenger properties [1]. Tan et al. [1] and Reiter et al. [5] induced hydroxyl radical formation by photolysis of a hydrogen peroxide solution under UV irradiation (254 nm) and showed that the concentration required to inhibit the hydroxyl radical formation by 50% was 6-fold and 14-fold lower for melatonin compared to glutathione and mannitol, respectively. Melatonin detoxifies reactive oxidants by donating an electron and forming an indoyl cation radical. In the presence of the superoxide anion radical it is immediately converted to the nonreactive kynuramine metabolite (N-acetyl-N-formyl-5-methoxykynuramine) [6]. Hydroxyl radicals are reported to be the most damaging ones, and melatonin quenches especially these radicals, but also others such as the singlet oxygen molecule [7, 8]. The antioxidative capacities of melatonin were later confirmed by other groups and were partially explained by the inhibition and deactivation of redox active

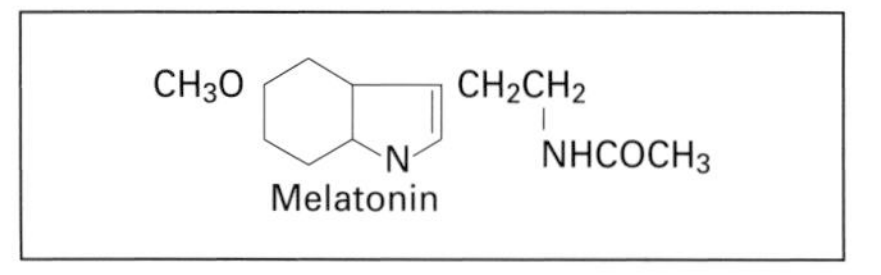

Fig. 1. Chemical structure of melatonin (N-acetyl-5-methoxytryptamine).

metal ions [9, 10]. Because melatonin scavenges mostly hydroxyl radicals which are produced under the influence of UV light, studies with melatonin were performed especially in the field of photodermatology.

In vitro Models

The antioxidant potential of melatonin concerning skin or UV influences on the skin was investigated in an in vitro model with IL-3-stimulated leukocytes [11]. Human leukocytes were taken from EDTA whole blood from healthy volunteers and put into small test-tubes. After isolation by density gradient with dextran, the supernatants were centrifuged for 15 min with 900 rpm, remaining erythrocytes were hemolyzed and then the cell suspensions were taken up in 30 ml IL-3-stimulated buffer solution. The cell number was counted and suspensions with a mean cell concentration of about 1.5×10^6/ml were used for further investigation. The IL-3-stimulated cell solutions were divided into 10 aliquots and distributed into UV-light-permeable quartz glass Petri dishes. To 2,250 μl cell solution, 250 μl of either melatonin or PBS was added to 5 aliquots, respectively. PBS served as a control. The melatonin stock solution contained 20 mmol melatonin and was diluted with acetonitrile (final concentration below 2%) and filled up with PBS. The final concentration of melatonin and PBS in the cell suspension of 2,500 μl was 2 mmol.

UV irradiation of the cell solutions resulted in an increasing formation of reactive oxygen species (ROS). A Waldmann UV800 UVB lamp (280–360 nm; maximum 310 nm) was used to apply UV doses between 75 and 750 mJ/cm^2. For investigation of cell viability, the cell solutions were irradiated with higher doses of UV light up to 2.25 J/cm^2.

The measurement of ROS was performed with the chemiluminescence technique with lucigenin using the luminometer LB 953 as described elsewhere [12]. To assess viability, 20 μl of each preparation was stained with 20 μl trypan blue, and viable cells were counted under the microscope.

It could be shown that leukocytes produce a respectable amount of ROS under the irradiation with UVB light which is proportional to the UV dose. Under irradiation with 75 mJ/cm UVB light, the radical formation in controls (PBS) raised up to twice higher levels above baseline ROS, whereby melatonin-

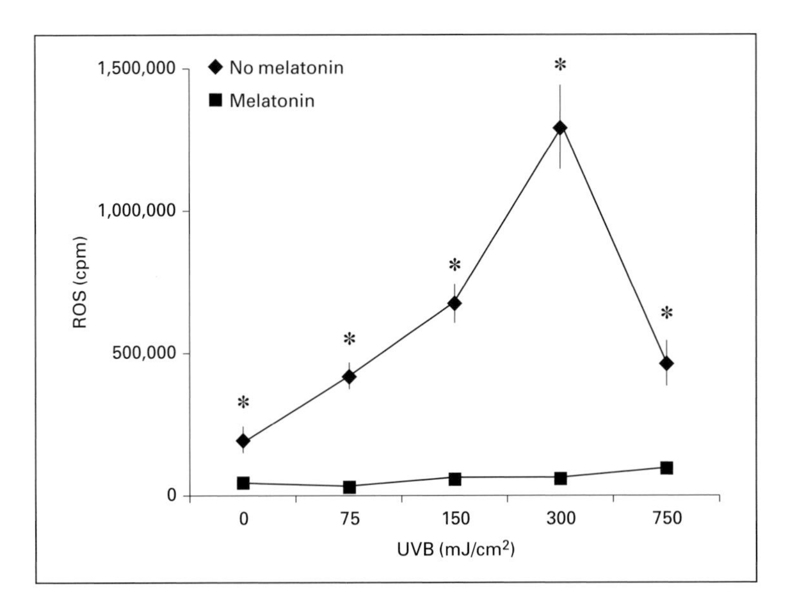

Fig. 2. Increasing UV doses led to an increase in ROS in untreated leukocyte solutions compared to low radical levels in the melatonin-treated cell solutions. A strong decrease was seen under irradiation with UVB at a dose of 750 mJ/cm^2. The viability in these cell solutions was decreased as doses above 300 mJ/cm^2 were cytotoxic. $^*p < 0.001$.

treated cells rested in the same range as the baseline value. Increasing UVB doses of 150 and 300 mJ/cm^2 led to a linear increase in the counts per minute for ROS in PBS solutions with a maximum at 300 mJ/cm^2 resulting in counts per minute of 1.2 Mio. In melatonin-treated cell solutions, the radical formation yielded only 48,000 cpm. Irradiation of the cell solutions with 750 mJ/cm showed a drastic decrease in radical formation in PBS solutions. In contrast, melatonin-treated leukocytes showed only a small continuous increase up to 80,000 cpm ROS formation under 750 mJ/cm^2 UVB irradiation (fig. 2).

The maximum ROS level in PBS-treated cell solutions was 13-fold higher than the baseline values, whereby maximum ROS values in melatonin-treated cell solutions were only 3-fold higher than unirradiated baseline values.

To investigate the dose dependency of melatonin under the influence of UVB irradiation, two different experimental designs were chosen: the first series was performed with concentrations of melatonin over a wide range of powers of 10 to find a maximum effect (0.1 nmol, 1 nmol, 10 nmol, 100 nmol, 1 μmol, 10 μmol, 100 μmol, 1 mmol; fig. 3). The second preparation was performed with melatonin in a small-range optimum, which was found in the wide-range studies of concentrations (0.1, 0.5, 1, 2, 3, 5, 7.5, 10 mmol; fig. 4).

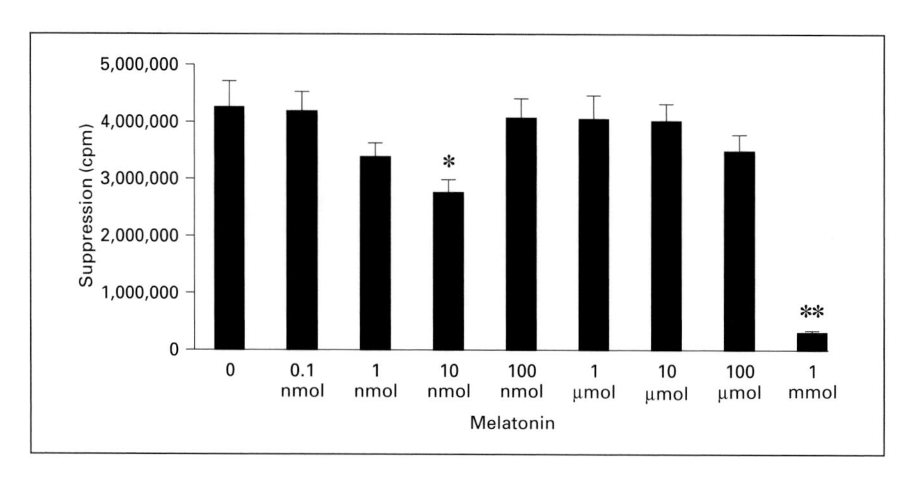

Fig. 3. Addition of melatonin to leukocyte solutions in a range from 0.1 nmol to 1 mmol under irradiation with UVB light at a dose of 750 mJ/cm^2 showed a biphasic effect of free radical suppression. Maximum suppression was visible at the concentration of 10 nmol (*$p<0.01$) and 1 mmol (**$p<0.001$).

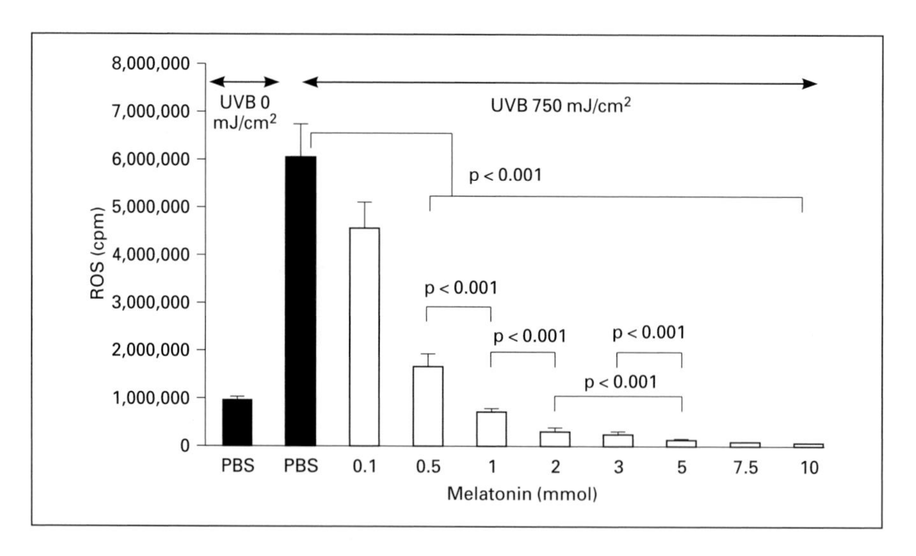

Fig. 4. Direct dose-response relationship between increasing melatonin concentrations and increase in free radical suppression under the influence of UVB light. A significant reduction of free radical formation was shown with melatonin concentrations of 0.5–10 mmol. Differences in this range were statistically relevant between 0.5 and 1 mmol, 1 and 2 mmol and between 2 and 3 versus 5 mmol melatonin.

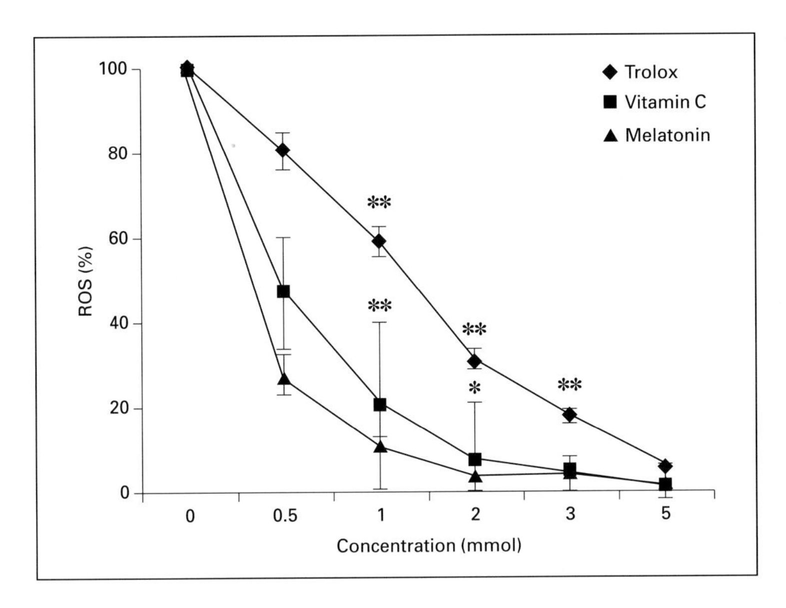

Fig. 5. Suppression of ROS by melatonin compared to vitamin C and trolox. Melatonin suppressed the radical formation significantly better than vitamin C and trolox at concentrations of 1, 2 and 3 mmol (*$p<0.05$; **$p<0.001$). Cell solutions were irradiated with UVB light at a dose of 750 mJ/cm^2.

In the wide-range concentrations there was a biphasic dose relationship concerning the suppression of free radicals under UVB light [13]. The maximum effect was shown in the concentration of 1 mmol melatonin ($p<0.001$) and a second maximum was seen in the concentration of 10 nmol ($p<0.01$) with a significant difference to the untreated UVB-irradiated cell solution. 1 mmol melatonin suppressed the radical formation by 17 times (fig. 3). In this range, a direct dose-effect relationship was found with increasing melatonin concentrations starting with 0.1 mmol melatonin up to 10 mmol. There was a significant suppression of ROS by concentrations of 0.5–10 mmol ($p<0.001$); 5 mmol led to a 60-fold and 10 mmol to a 260-fold suppression of radical formation under the influence of UVB light (fig. 4).

In the same model, melatonin was compared with a vitamin E analog (trolox) and glycolic acid (vitamin C) [14]. Formation of ROS in leukocyte solutions treated with trolox at a concentration of 0.5 mmol suppressed the radical formation down to 80%. Suppression of 50% was reached at a concentration of 1.3 mmol trolox (fig. 5). Vitamin C in a concentration of 1.1 mmol led to a suppression of radical formation of 80%; 0.5 mmol already reduced the counts per minute by 50%. Melatonin treatment showed a significant reduction of radical formation compared to vitamin C and trolox.

Melatonin was the strongest suppressor of ROS in this model under UVB irradiation and revealed a more than 2-fold effect compared to vitamin E in all concentrations except that of 5 mmol. In the concentrations between 1 and 3 mmol, the difference was statistically significant ($p<0.001$). Compared to vitamin C, melatonin showed a significantly stronger suppression of radicals in the concentration of 1 mmol ($p<0.001$) and 2 mmol ($p<0.05$). Even though in other studies vitamin E was shown to be a strong radical scavenger and superior to melatonin, it could be demonstrated that melatonin seems to be the most effective radical scavenger in IL-3-stimulated leukocytes under UVB irradiation.

Pieri et al. [9] showed similar effects in a model with 2-2′-azo-bis(2-amidino-propane)dihydrochloride as a peroxyl radical generator. The radical scavenger ability was expressed as the oxygen-radical-absorbing capacity units indicated by β-phycoerythrin as fluorescent indicator protein. Melatonin revealed a twice higher activity concerning the scavenging properties of free radicals compared to trolox [9].

In a model using lipid bilayers and human skin homogenates, the antioxidant activity of melatonin and vitamin E (α-tocopherol) in combination with α-glycolic acid was compared. The antioxidative capacity of the substances was assessed by detecting the lipid peroxidation spectrophotometrically by the time course of lipid hydroperoxide production in liposomes and by formation of thiobarbituric-acid-reactive substances in skin homogenates. In peroxidation of liposomes, glycolic acid showed a synergistic effect of 250 and 80% with vitamin E and melatonin, respectively. Even though in this model melatonin was not as potent as vitamin E, an antioxidative effect could be shown and melatonin was able to be recycled by glycolic acid [15].

Clinical Studies

In clinical studies, it could be shown that melatonin had a suppressive effect on UV erythema in humans. One study was performed with 20 healthy volunteers which were irradiated with UVB light at a dose of 0.099 J/cm^2 on 4 defined skin areas on the lower back [16]. Immediately after the irradiation, the areas were treated with a nanocolloid gel containing melatonin in concentrations of 0.05, 0.1 and 0.5% or with carrier alone. The resulting erythema was evaluated 8 and 24 h after the UV exposure. The erythema measured by chromametry, a noninvasive biophysical method [17], differed significantly in the fields treated with 0.5% melatonin from the fields which were treated with 0.05% or with the carrier alone ($p<0.05$). Twenty-four hours after irradiation, the differences were no longer significant, but the study indicates clearly that

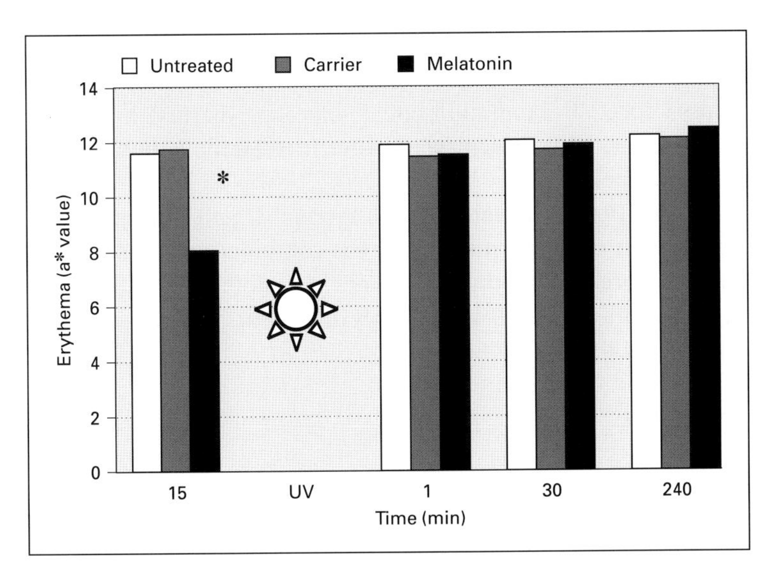

Fig. 6. Chromameter a* value 24 h after irradiation with UV light. The melatonin preparation applied 15 min before the irradiation led to a significantly lower erythema compared to carrier or untreated fields or to the fields treated after the irradiation. *$p < 0.001$.

melatonin had a protective effect concerning the UV-induced erythema in the human skin.

A second study was performed to evaluate eventual effects on the UV erythema dependent on the application time point. Melatonin was prepared in the most effective concentration of 0.5% of the preliminary study and compared with carrier alone or with untreated fields on the lower back of 20 healthy volunteers. The application of the topical preparations was performed 15 min before and 1, 30 and 240 min after the irradiation with UVB light at a dose of twice the MED. The UV erythema was assessed 24 h after irradiation and it could be shown that the skin treated with melatonin 15 min before irradiation developed a significantly weaker erythema compared to the carrier or the untreated skin at the same time point or the skin treated 1, 30 or 240 min after the irradiation (fig. 6) [18]. Melatonin could not exert a protective effect applied after irradiation.

This was explained by the hypothesis that melatonin has to be present in the skin at that time point when the acute irradiation occurs and that the time to penetrate the skin is about 15 min. The mechanism of action was explained with the direct radical-scavenging properties of melatonin and the neutralization of free radicals which are induced by UV light in the skin [19, 20]. Another mechanism was discussed depending on the modulation of proinflammatory mediators of arachidonic acid such as leukotrienes and prostaglandins which are produced under UV irradiation [21]. The results of the cell model with leukocytes exposed to UVB irradiation could underline the first hypothesis that the UV-protective quality of melatonin is at least partially based on its radical-scavenging properties.

Recently, combinations of melatonin with vitamin C and vitamin E were evaluated in humans. It could be shown that the combination containing all three components is more effective than melatonin in combination with vitamin C or vitamin E alone and that it is probably due to its radical scavenger properties [22].

Conclusions

Melatonin is a potent radical scavenger quenching especially hydroxyl radicals which are generated in the skin by UV irradiation [19, 23, 24]. Because melatonin is produced in the human body by the pineal gland, this fact may represent an endogenous protective mechanism against the UV-induced oxidative damage in the skin. The studies performed so far relate to the acute UV reaction in the skin, skin homogenates or in cell systems. Many studies with other antioxidants in photodermatology were conducted in a variety of in vitro systems. Vitamin E or its analogs were investigated in human keratinocytes [25], lipid bilayers and in skin homogenates [15]. Vitamin C was used alone or in combination with other antioxidants as a radical scavenger in porcine skin [26] and murine skin homogenates [27]. The antioxidant properties of melatonin and the above-mentioned antioxidants were compared in the leukocyte model under UV irradiation [14]. Since melatonin is also supposed to have antiaging properties [5, 28] which may be caused by immunomodulatory functions and biorhythmic regulatory circles or by the antioxidant activity in the human, other models such as the hairless mouse model seem preferable to study the possible effect of melatonin onto the skin aging process. Melatonin may act as a protective substance in intrinsic aging or in UV-induced extrinsic aging. There is some evidence that at least in the latter melatonin plays a major role.

References

1 Tan DX, Chen LD, Poeggeler B, Manchester LC, Reiter RJ: Melatonin: A potent, endogenous hydroxyl radical scavenger. Endocr J 1993;1:57–60.

2 Arendt J: Melatonin. Clin Endocrinol 1988;29:205–229.

3 Barrett P, Morris M, Choi WS, Ross A, Morgan PJ: Melatonin receptors and signal transduction mechanisms. Biol Signals Recept 1999;8:6–14.

4 Reiter RJ, Poeggeler B, Tan DX, Chen LD, Manchester LC, Guerrero JM: Antioxidant capacity of melatonin: A novel action not requiring a receptor. Neuroendocrinol Lett 1993;15:103–116.
5 Reiter RJ, Tan DX, Poeggeler B, Menendez-Pelaez A, Chen LD, Saarela S: Melatonin as a free radical scavenger: Implication for aging and age-related diseases. Ann NY Acad Sci 1994;719: 1–12.
6 Poeggeler B, Saarela S, Reiter RJ, Tan DX, Chen LD, Manchester LC, Barlow-Walden LR: Melatonin – A highly potent endogenous radical scavenger and electron donor: New aspects of the oxidation chemistry of this indole accessed in vitro. Ann NY Acad Sci 1994;738:49–420.
7 Halliwell B: Reactive oxygen species and the central nervous system. J Neurochem 1992;59:1609–1623.
8 Cagnoli CM, Atabay C, Kharlamova E, Manev H: Melatonin protects neurons from singlet oxygen-induced apoptosis. J Pineal Res 1995;18:222–226.
9 Pieri C, Marra M, Moroni F, Recchioni R, Marcheselli F: Melatonin: A peroxyl radical scavenger more effective than vitamin E. Life Sci 1994;55:271–276.
10 Princ FG, Maxit AG, Cardalda C, Batlle A, Juknat AA: In vivo protection by melatonin against delta-aminolevulinic acid-induced oxidative damage and its antioxidant effect on the activity of haem enzymes. J Pineal Res 1998;24:1–8.
11 Fischer T, Scholz G, Knöll B, Hipler UC, Elsner P: Effect of melatonin as a radical scavenger on UV-irradiated, IL-3-stimulated leucocytes. Arch Dermatol Res 1999;291:141.
12 Hipler UC, Schreiber G, Wollina U: Reactive oxygen species in human semen: Investigations and measurements. Arch Androl 1998;40:67–78.
13 Fischer T, Scholz G, Knöll B, Hipler UC, Elsner P: Dose-response of melatonin (N-acetyl-5-methoxytryptamine) as a radical scavenger on UV-irradiated, IL-3-stimulated leucocytes. J Invest Dermatol 1999;112:610.
14 Fischer TW, Scholz G, Knöll B, Hipler UC, Elsner P: Efficacy of melatonin compared to vitamin C and trolox as radical scavengers in UVB-irradiated, IL-3-stimulated leucocytes. Arch Dermatol Res 2000;292:101.
15 Morreale M, Livrea MA: Synergistic effect of glycolic acid on the antioxidant activity of alpha-tocopherol and melatonin in lipid bilayers and in human skin homogenates. Biochem Mol Biol Int 1997;42:1093–1102.
16 Bangha E, Elsner P, Kistler GS: Suppression of UV-induced erythema by topical treatment with melatonin (N-acetyl-5-methoxytryptamine): A dose response study. Arch Dermatol Res 1996;288: 522–526.
17 Elsner P: Chromametry: Hardware measuring principles and standardization of measurements; in Berardesca E, Elsner P, Maibach HI (eds): Handbooks of Skin Bioengineering: Cutaneous Bloodflow and Erythema. Boca Raton, CRC Press, 1994, pp 247–252.
18 Bangha E, Elsner P, Kistler GS: Suppression of UV-induced erythema by topical treatment with melatonin (N-acetyl-5-methoxytryptamine): Influence of the application time point. Dermatology 1997;195:248–252.
19 Jurkiewicz BA, Buettner GR: Ultraviolet light-induced free radical formation in skin: An electron paramagnetic resonance study. Photochem Photobiol 1994;59:1–4.
20 Taira J, Mimura K, Yoneya T, Hagi A, Murakami A, Makino K: Hydroxyl radical formation by UV-irradiated epidermal cells. J Biochem 1992;111:693–695.
21 Punnonen K, Puustinen T, Jansen CT: Ultraviolet B irradiation induces changes in the distribution and release of arachidonic acid, dihomo-gamma-linolenic acid, and eicosapentaenoic acid in human keratinocytes in culture. J Invest Dermatol 1987;88:611–614.
22 Dreher F, Gabard B, Schwindt DA, Maibach HI: Topical melatonin in combination with vitamins E and C protects skin from ultraviolet-induced erythema: A human study in vivo. Br J Dermatol 1998;139:332–339.
23 Masaki H, Atsumi T, Sakurai H: Detection of hydrogen peroxide and hydroxyl radicals in murine skin fibroblasts under UVB irradiation. Biochem Biophys Res Commun 1995;206:474–479.
24 Shindo Y, Witt E, Han D, Packer L: Dose-response effects of acute ultraviolet irradiation on antioxidants and molecular markers of oxidation in murine epidermis and dermis. J Invest Dermatol 1994;102:470–475.

25 Malorni W, Straface E, Donelli G, Giacomoni PU: UV-induced cytoskeletal damage, surface blebbing and apoptosis are hindered by alpha-tocopherol in cultured human keratinocytes. Eur J Dermatol 1996;6:414–420.
26 Darr D, Combs S, Dunston S, Manning T, Pinnell S: Topical vitamin C protects porcine skin from ultraviolet radiation-induced damage. Br J Dermatol 1992;127:247–253.
27 Kitazawa M, Podda M, Thiele J, Traber MG, Iwasaki K, Sakamoto K, Packer L: Interactions between vitamin E homologues and ascorbate free radicals in murine skin homogenates irradiated with ultraviolet light. Photochem Photobiol 1997;65:355–365.
28 Lesnikov VA, Pierpaoli W: Pineal cross-transplantation (old-to-young and vice versa) evidence for an endogenous 'aging clock'. Ann NY Acad Sci 1994;719:456–460.

Dr. Tobias W. Fischer, Department of Dermatology and Allergy,
Friedrich-Schiller-Universität Jena, Erfurter Strasse 35, D–07740 Jena (Germany)
Tel. +49 3641 937320, Fax +49 3641 937430, E-Mail tobias.fischer@med-uni.jena.de

Thiele J, Elsner P (eds): Oxidants and Antioxidants in Cutaneous Biology.
Curr Probl Dermatol. Basel, Karger, 2001, vol 29, pp 175–186

Bioconversion of Vitamin E Acetate in Human Skin

Zeenat Nabi [a], *Amir Tavakkol* [a], *Marek Dobke* [b], *Thomas G. Polefka* [a]

[a] Colgate-Palmolive Co., Piscataway, N.J., and
[b] Department of Surgery, University of Medicine and Dentistry of New Jersey, Newark, N.J., USA

Due to its constant exposure to various environmental factors (i.e. UV radiation, air pollution and other oxidants), the skin is susceptible to oxidative stress. Oxidative stress occurs when the prooxidant species exceed the antioxidant species. For the skin, the consequences of this imbalance include inflammation, phototoxicity, accelerated aging, degradation of function and skin cancer [1, 2].

As the first line of defense against noxious environmental factors, nature has endowed the epidermis (outermost layer of skin) with a sophisticated antioxidant defense system [3]. Although ascorbic acid (vitamin C) is the most prevalent antioxidant in the epidermis [4], some scientists believe that vitamin E (tocopherol) is the most important antioxidant. This belief is based on the following facts: vitamin E is present in limited quantities; it is readily depleted by UV radiation [5] and other oxidative stresses [6], and it is a key antioxidant for preserving the most important function of the skin, namely the barrier function [7]. Vitamin E accomplishes this feat by protecting and limiting oxidation of the stratum corneum polyunsaturated lipids [8].

In addition to its barrier function, the skin serves as a portal of entry for topical drugs and cosmetics. Although frequently referred to as a layer of dead cells, the stratum corneum exhibits considerable enzymatic activity [9, 10]. Interestingly, these enzymes not only act upon their natural substrates but in some cases can modify xenobiotics. Indeed, the skin's capacity to transform inactive compounds to 'activated' species is the basis of 'prodrug' therapeutics. Prodrugs are defined as inactive compounds that are transformed into active species (by enzymes or other means) at or near their site of action.

There is substantial interest in delivering the antioxidant benefits of vitamin E (tocopherol) to the skin [for reviews, see 7, 11]. Unfortunately, the sensitivity of vitamin E to oxidation [8] has limited its use in topical personal care products. To overcome this deficiency, formulators of topical products frequently use the stable esters of vitamin E (e.g. acetate, linolenate, sorbate and succinate). This approach is based on the premise that enzymes located in the skin will transform the provitamin E to the active vitamin E. Although evidence supporting this premise is well established in animals [12–20], results from human studies are scarce and conflicting. Cesarini et al. [21] and Montenegro et al. [22] reported that topical application of provitamin E (tocopheryl acetate) to human volunteers reduced UV-induced sunburn cell formation and erythema, respectively. Clearly, these data imply that vitamin E acetate was transformed to the active vitamin E. However, in a recent human clinical study, Alberts et al. [23] did not observe conversion of vitamin E acetate to vitamin E, although the topically applied vitamin E acetate was substantially absorbed by the skin. These investigators concluded that vitamin E acetate is not metabolized to vitamin E in the skin [23].

In this report we use three different systems, namely living skin equivalents (LSE™), EpiDerm™ and freshly excised human skin, to demonstrate the capacity of human skin to bioconvert vitamin E acetate to vitamin E. We also show that this process is enzyme dependent since it can be partially inhibited by the well-known esterase inhibitor diisopropyl fluorophosphate (DFP). Lastly, we demonstrate that vitamin E generated from the bioconversion of vitamin E acetate can neutralize reactive oxygen species, e.g. cumene hydroperoxide.

Materials and Methods

Reagents

All solutions and reagents were of analytical grade and obtained from Sigma Chemical Co. (St. Louis, Mo., USA) and Aldrich Chemical Co (Milwaukee, Wisc., USA). MTT (thiazolyl blue), DFP and cumene hydroperoxide were all from Sigma Chemical Co. Vitamin E (tocopherol) and vitamin E acetate (tocopheryl acetate) were purchased from Fluka Chemical Corporations (Ronkonkoma, N.Y., USA). All culture media, serum and buffers including phosphate-buffered saline (PBS) were purchased from Gibco BRL (Grand Island, N.Y., USA), unless otherwise stated. The human LSE ZK1301 was purchased from Advanced Tissue Sciences (La Jolla, Calif., USA) and the human EpiDerm skin model was from the MatTek Corporation (Ashland, Mass., USA). Human skin (explants) were obtained from patients undergoing breast reduction surgery at the University of Medicine and Dentistry of New Jersey. All patients had signed an Institutional Review Board (IRB) approved consent form prior to surgery.

Organ Culture Maintenance

Upon arrival, EpiDerm and LSE model skin samples were cultured overnight in the manufacturer's recommended vitamin-E-free maintenance media. Explants were rinsed thoroughly with PBS, sliced into pieces (approx. 0.36 cm^2) and transferred onto Millicells® supports. The tissue pieces were rinsed with RPMI medium containing 5% fetal calf serum and an antibiotic/fungicide cocktail, and maintained at the air-liquid interface in the same medium. A color change in the medium's pH indicator was taken as evidence of tissue viability. Skin explants were used within 2–3 days of surgery. All cultures were maintained in an incubator at 37 °C with a humidified atmosphere of 5% CO_2. Tissue viability was measured by the MTT assay according to Mosmann [24].

Topical Treatment and Vitamin E Analysis

Human skin equivalents (LSE and EpiDerm) and skin explants were treated topically with 4 mg/cm^2 of an oil/water emulsion containing 1% vitamin E acetate. Model skin samples were incubated at 37 °C for varying time points in their respective maintenance media. Following treatment, skin samples were homogenized and extracted (LSE) or extracted (EpiDerm and skin explant) in ice-cold ethanol (1 ml) for 1 h at 4 °C. The samples were centrifuged and the supernatant was transferred to another vial. After a second extraction procedure, the ethanol extracts were pooled, evaporated to dryness under N_2 and redissolved in 0.5 ml methanol:isopropanol:butanol (70:20:10). Vitamin E and vitamin E acetate were quantified by HPLC using an Ultracarb™ 5µ-ODS 20 column (100 × 4.6 mm; Phenomenex®, Torrance, Calif., USA) and eluted with a mobile phase composed of methanol:isopropanol: butanol (75:20:5) at a flow rate of 0.75 ml/min. The vitamins were detected with a Waters UV-Vis model 484 detector (Millipore, Milford, Mass., USA) set to 288 nm. Under these conditions, the typical retention times for vitamin E and vitamin E acetate were 5 and 7 min, respectively. The involvement of skin esterase(s) in the hydrolysis of vitamin E acetate to vitamin E was determined by prior treatment of EpiDerm skin with varying concentrations of DFP, an esterase inhibitor [25], as described in the Results section.

To study the protective effect of vitamin E generated by the bioconversion of vitamin E acetate, EpiDerm samples were pretreated topically for 30 min with an oil/water emulsion ± vitamin E acetate (1%, w/w). After 30 min, the emulsion was washed off and the tissue challenged with various doses of cumene hydroperoxide (in 50 µl equal parts of ethanol:PBS) for 30 min. The tissues were returned to the incubator, and tissue viability was determined by the MTT test 24 h later.

Statistical Analysis

All results are represented as means ± standard deviation. Statistical comparisons were made using Student's t test.

Results

To study the metabolism of vitamin E in skin, we used three human skin models: two skin equivalents, LSE and EpiDerm, and skin derived from cosmetic surgery. The LSE is a three-dimensional skin model consisting of several layers of proliferating and metabolically active human neonatal foreskin-derived dermal fibroblasts and epidermal keratinocytes grown on a nylon mesh (fig. 1a). Despite its original popularity, the LSE is no longer available for basic research. However, another model, the EpiDerm, has become available. The EpiDerm model consists of human epidermal keratinocytes cultured on membrane filters to form multilayered stratified epidermis. As can be seen in figures 1a and b, the EpiDerm differs from the LSE in that it lacks a viable dermis. Figure 1c is a photomicrograph of full-thickness human skin derived from cosmetic surgery. Note the well-defined boundaries between the stratum corneum, epidermis and dermis.

Bioconversion of Vitamin E Acetate to Vitamin E

Model Skin Substrates LSE and EpiDerm

The ability of human skin to bioconvert vitamin E acetate to vitamin E was initially examined in the LSE model. An oil/water emulsion containing vitamin E acetate (1%, w/w) was applied topically (4 mg/cm^2) to the skin samples which were then incubated from 0 to 14 h at 37 °C. The skin samples were extracted with ethanol, and vitamin E was measured by HPLC as described in Methods. Figure 2 shows the time course for the bioconversion of vitamin E acetate to vitamin E in the LSE model. Vitamin E was detectable as early as 60 min (0.07 nmol/cm^2) and rose dramatically to 0.43 nmol/cm^2 skin at 6 h; attaining a maximum value of 0.50 nmol/cm^2 at 14 h. There were no detectable levels of vitamin E in untreated skin.

To better localize the site of bioconversion, we chose to use the EpiDerm skin model. Unlike the LSE, the EpiDerm model contains only an epidermis. In this experiment, the skin model was treated as before and vitamin E measured at 6 and 12 h after treatment. The results shown in figure 3 confirm the capacity of epidermal keratinocytes to bioconvert vitamin E acetate to vitamin E.

Excised Human Skin (Explant)

Although LSE and EpiDerm exhibit many structural and biochemical similarities to human skin, their stratum corneum does not provide the same quality of barrier found in vivo [26–28]. Therefore, we decided to use skin explant derived from mammoplasty. Skin explants (0.36-cm^2 pieces) were treated topically with an oil/water emulsion with and without vitamin E acetate

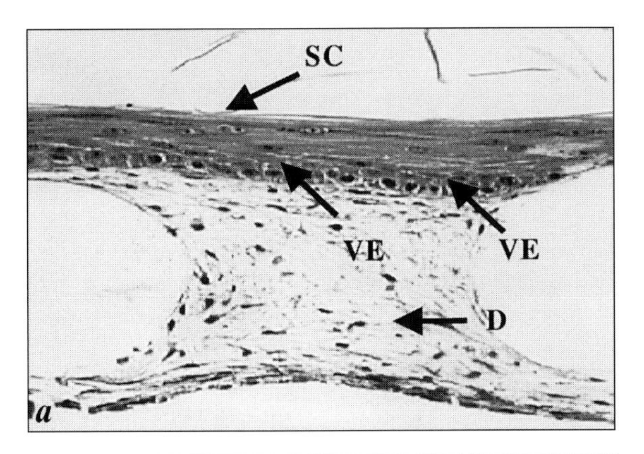

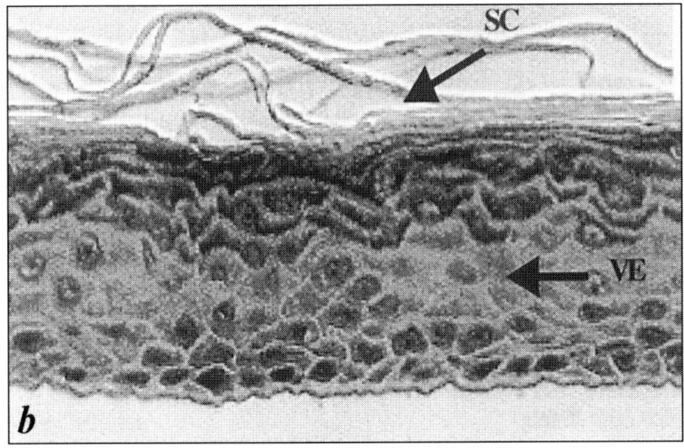

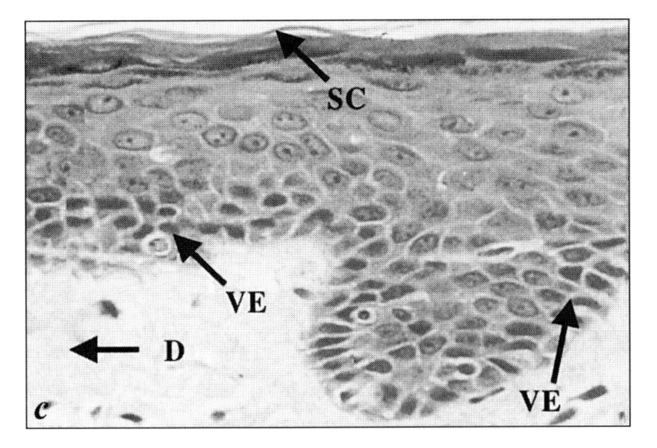

Fig. 1. Photomicrographs of cross-sections of three human skin systems: LSE (***a***), EpiDerm (***b***) and human skin (***c***). Note that the EpiDerm lacks a dermis. In **a–c**, arrows show stratum corneum (SC), viable epidermis (VE) and dermis (D).

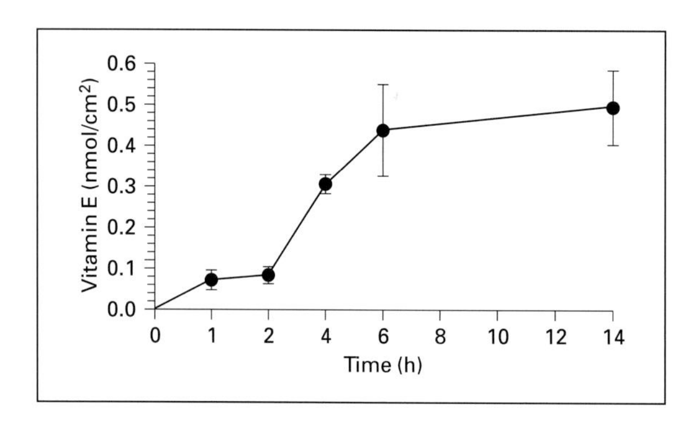

Fig. 2. Bioconversion of vitamin E acetate to vitamin E in LSE. The LSE samples were treated with an oil/water emulsion containing vitamin E acetate for 0–14 h. Treated samples (n = 3) were extracted with ethyl alcohol, and vitamin E was analyzed by HPLC. Each data point represents the mean ± SD.

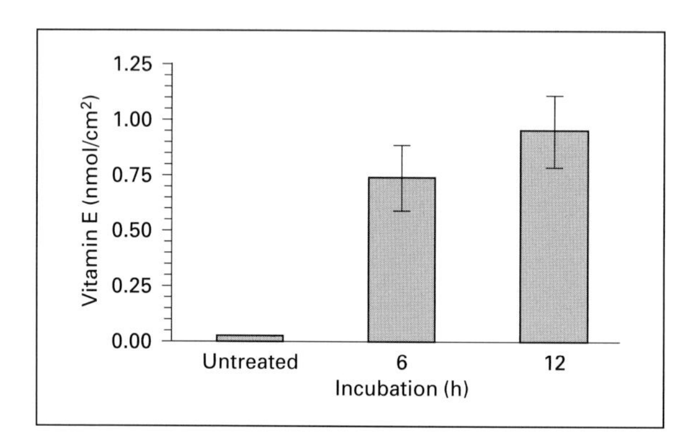

Fig. 3. Bioconversion of vitamin E acetate in EpiDerm skin. The EpiDerm skin was treated with an oil/water emulsion containing vitamin E acetate and incubated for 6 and 12 h. Treated and untreated samples (n = 3) were extracted and analyzed as in figure 2. Each data point represents the mean ± SD.

(1%, w/w) for up to 21 h. Figure 4 clearly shows that vitamin E in skin increases with contact time, establishing the capacity of human skin to bioconvert vitamin E acetate to vitamin E.

Vitamin E Acetate Is Metabolized to Vitamin E by Skin Esterases

The results reported above do not exclude the nonspecific hydrolysis of the vitamin E ester in the epidermis. To determine whether the bioconversion

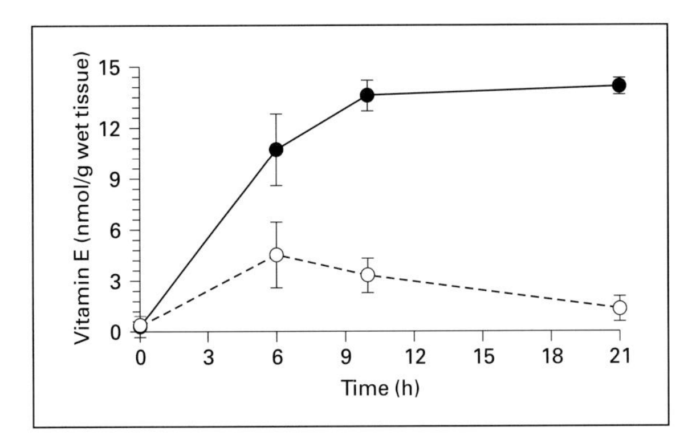

Fig. 4. Bioconversion of vitamin E acetate in human skin explant. Excised human skin samples were treated with an oil/water emulsion containing vitamin E acetate and incubated for 0–21 h. Treated samples (n = 3) were extracted with ethyl alcohol, and vitamin E was analyzed (cf. Methods). Each data point represents the mean ± SD. Vitamin E levels were significantly higher in vitamin-E-acetate-treated skin compared to vehicle-treated skin (6 h, p = 0.019; 10 h, p = 0.0001, and 21 h, p = 0.0001). The solid line represents the vitamin-E-acetate-treated skin and the dashed line the vehicle-treated skin.

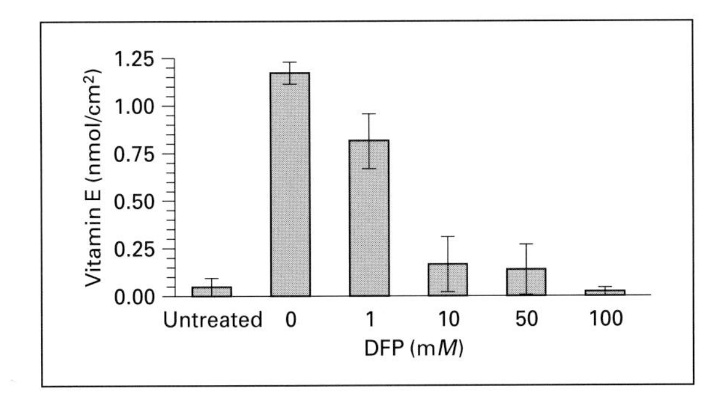

Fig. 5. Vitamin E acetate metabolism is inhibited by DFP. Human skin explant was pretreated with 0.0, 1.0, 10, 50 and 100 mM DFP for 30 min. The oil/water emulsion containing vitamin E acetate was applied to the skin (n = 4), which was then incubated for 6 h at 37 °C. Vitamin E formation was determined by HPLC as before. Each data point represents the mean ± SD.

was mediated by enzymes (e.g. esterases), the human skin explant was pretreated for 30 min with DFP prior to application of vitamin E acetate. Bando et al. [25] have shown that DFP is a lipophilic, nonspecific esterase inhibitor. Figure 5 shows that the enzyme-mediated bioconversion of vitamin E acetate to vitamin E is sensitive to DFP. As can be seen, exposure to 100 m*M* DFP

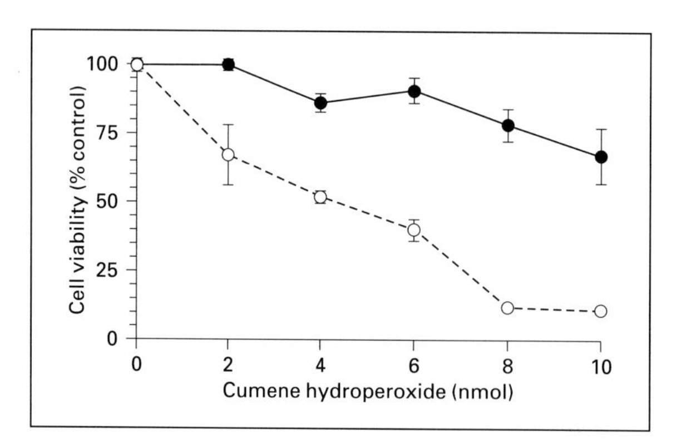

Fig. 6. Vitamin E acetate reduces peroxide-induced skin cell damage. EpiDerm skin (n = 3) was treated with an oil/water emulsion containing vitamin E acetate for 30 min and challenged with various concentrations of cumene hydroperoxide for an additional 30 min. The antioxidant efficacy of bioconverted vitamin E was determined by measuring cell viability by the MTT test at 24 h. Each data point represents the mean ± SD. Percent cell viability was significantly higher in vitamin-E-acetate-treated skin at all concentrations of cumene hydroperoxide used ($p \leq 0.05$). The solid line represents the vitamin-E-acetate-treated skin and the dashed line the vehicle-treated skin.

for 30 min almost completely inhibits bioconversion. These data strongly suggest that vitamin E acetate bioconversion in skin is mediated by esterases. Similar results were obtained using the Epiderm model [29].

Protective Efficacy of Bioconverted Vitamin E

The results reported thus far establish the ability of human skin to bioconvert vitamin E acetate to vitamin E. To establish the antioxidant efficacy of vitamin E resulting from the bioconversion of vitamin E acetate, EpiDerm skin was pretreated with an oil/water emulsion ± vitamin E acetate (1%, w/w). After 30 min, the emulsion was washed off, and the tissue was challenged with various doses of cumene hydroperoxide for 30 min. Twenty-four hours later, cell viability was determined by the MTT test (cf. Methods). Figure 6 shows that the skin pretreated with vitamin E acetate was better protected than the skin treated with the vehicle at each dose of cumene hydroperoxide. Indeed, skin pretreated with the vitamin E acetate emulsion and then exposed to 10 nmol/cm^2 cumene hydroperoxide retained greater than 60% viability. In contrast to this, the viability of the control skin, pretreated with the vehicle emulsion, was completely lost.

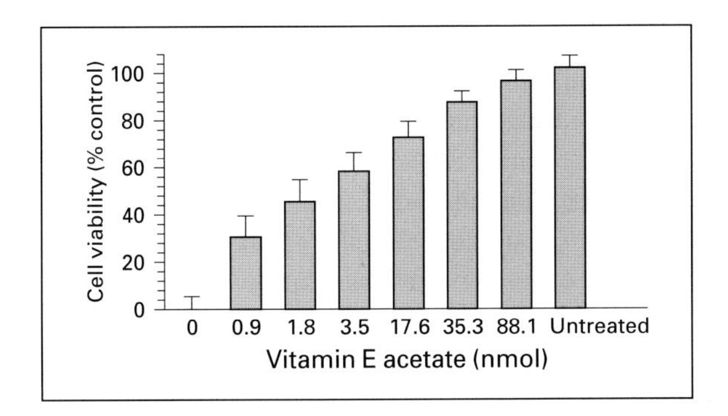

Fig. 7. Vitamin E acetate protects the skin in a dose-dependent fashion. Varying amounts of vitamin E acetate (in an oil/water emulsion) were applied to EpiDerm for 3 h. Skin samples ($n = 3$) were then challenged with 2 nmol cumene hydroperoxide for 30 min. They were then rinsed with PBS and incubated for 24 h in the maintenance medium. Cell viability was analyzed as in figure 6. Each data point represents the mean $\pm$ SD. For all points $p < 0.0001$.

We next determined the protective dose of vitamin E acetate necessary to mitigate the deleterious effects of cumene hydroperoxide. The EpiDerm skin model was pretreated topically for 3 h with an oil/water emulsion containing various doses of vitamin E acetate. The skin samples were then challenged with 2 nmol cumene hydroperoxide for 30 min, rinsed with PBS and cultured for 24 h in the maintenance medium. Tissue viability was measured by the MTT assay. The results shown in figure 7 reveal that the application of vitamin E acetate to the EpiDerm skin substantially reduced the cytotoxic effects of cumene hydroperoxide. A dose response relationship was observed between 0.9 and 35 nmol of vitamin E acetate, with maximum protection (92.7%) observed at 88 nmol of vitamin E acetate. Under these experimental conditions (i.e. 3 h pretreatment with vitamin E acetate followed by 30 min exposure to 2 nmol cumene hydroperoxide), 2.3 nmol vitamin E acetate would be required to prevent 50% loss of viability.

Taken together, the results presented in figures 6 and 7 show that human skin is capable of converting inactive vitamin E acetate to the biologically active antioxidant. Moreover, the resulting antioxidant is capable of protecting the tissue from an oxidative challenge.

Discussion

In this report we have demonstrated, for the first time, that human skin possesses the enzymes necessary to bioconvert topically applied vitamin E acetate (tocopheryl acetate) to vitamin E (tocopherol). In animals, evidence supporting this bioconversion is well established [12–20]. However, in human skin, such evidence is scarce and conflicting. Although Cesarini et al. [21] and Montenegro et al. [22] provided indirect clinical evidence supporting bioconversion, Alberts et al. [23] were unable to detect hydrolysis of topically applied vitamin E acetate. Thus, these latter investigators concluded that human skin lacks the ability to metabolize vitamin E acetate to vitamin E.

To study the metabolism of vitamin E acetate in human skin, we employed three model systems (fig. 1). The two living skin equivalent models were chosen because of their structural, biochemical and functional (i.e. permeability barrier) similarities with human skin in vivo [26–28]. The third model, human skin explant, is an ex vivo model derived from human subjects undergoing cosmetic surgery. This latter model represents the closest analog to human skin in vivo [29].

Our results show that the lipophilic vitamin E acetate is adsorbed readily into the skin where it is hydrolyzed to vitamin E. Metabolism of the provitamin was observed in all three models. Surprisingly, rates of hydrolysis were consistent among all models (LSE = 0.07; EpiDerm = 0.12; explant = 0.18 nmol/h/cm^2). Since the EpiDerm lacks a dermis, the results suggest that a dermis is not necessary for bioconversion. Tauber and Rost [30] obtained similar results; specifically, hydrolysis of corticosteroid 21 esters was approximately 20-fold greater in the epidermis than in the dermis. Taken together, we believe that the stratified epidermis is the dominant site for the hydrolysis of vitamin E acetate.

Although the bioconversion of vitamin E acetate to vitamin E has been shown in animal skin [12–20], few investigators have provided data to explain how this hydrolysis is mediated. Without proof, most investigators have attributed this hydrolysis to skin enzymes. Using the human skin explant model, we show for the first time that hydrolysis of vitamin E acetate can be inhibited by DFP, a nonspecific esterase inhibitor (fig. 5). Additionally, this inhibition is dose dependent. In our opinion, these inhibition studies are currently the strongest evidence supporting our contention that human skin has the capacity to bioconvert vitamin E acetate to vitamin E. Moreover, we show that the vitamin E generated from the hydrolysis of vitamin E acetate mitigates the effects of chemically induced oxidative stress (fig. 6, 7).

To maintain the health and beauty of their skin, consumers are interested in the therapeutic benefits of vitamin E. Since vitamin E is readily oxidized, vitamin E acetate is frequently used in skin care products based on the premise that enzymes in the skin will bioconvert vitamin E acetate to vitamin E. In contrast to Alberts et al. [23], our studies extend the work done in animals by showing that human skin also has the capacity to bioconvert vitamin E acetate to vitamin E. Additionally, the results presented here suggest that hydrolysis of vitamin E acetate is mediated by an esterase that is susceptible to inhibition.

Acknowledgements

We are indebted to N. Soliman for vitamin E formulations, Dr. J. Mattai for developing the HPLC methodology for vitamin E analysis and P. Stravalexis, CCM (UMDNJ), for providing excised human skin. The authors also acknowledge S. Cardona and B. Wolff for excellent technical assistance and Dr. M. Kotler for the statistical analysis of the data. We also thank Mr. J. Sheasgreen (MatTek Corporation) for his advice on the use of EpiDerm skin.

References

1 Witt EH, Motchnik P, Packer L: Evidence for UV light as an oxidative stressor in skin; in Fuchs J, Packer L (eds): Oxidative Stress in Dermatology. New York, Dekker, 1993, pp 29–47.
2 Emerit I: Free radicals and aging of the skin; in Emerit I, Chance B (eds): Free Radicals and Aging. Basel, Birkhäuser, 1992, pp 328–341.
3 Fuchs J, Mehlhorn RJ, Packer L: Free radical reduction mechanisms in mouse epidermis skin homogenates. J Invest Dermatol 1989;93:633–640.
4 Shindo Y, Witt E, Han D, Packer L: Dose response effects of acute ultraviolet irradiation on antioxidants and molecular markers of oxidation on murine epidermis and dermis. J Invest Dermatol 1994;102:470–475.
5 Thiele JJ, Traber MG, Packer L: Depletion of human stratum corneum vitamin E: An early and sensitive in vivo marker of UV photo-oxidation. J Invest Dermatol 1998;110:756–761.
6 Thiele JJ, Traber MG, Polefka TG, Cross CE, Packer L: Ozone-exposure depletes vitamin E and induces lipid peroxidation in murine stratum corneum. J Invest Dermatol 1997;108:753–757.
7 Mayer P: The effects of vitamin E on the skin. Cosmet Toiletries 1993;108:99–114.
8 Kamal-Eldin A, Appelqvist LA: The chemistry and antioxidant properties of tocopherols and tocotrienols. Lipids 1996;31:671–701.
9 Forestier JP: Les enzymes de l'espace extra-cellulaire du stratum corneum. Int J Cosmet Sci 1992; 14:47–63.
10 Redoules D, Tarroux R, Perie J: Epidermal enzymes: Their role in homeostasis and their relationships with dermatosis. Skin Pharmacol Appl Skin Physiol 1998;11:183–192.
11 Rangarajan M, Zatz JL: Skin delivery of vitamin E. J Cosmet Sci 1999;50:249–279.
12 Bisset DL, Chatterjee R, Hannon DP: Photoprotective effect of superoxide-scavenging antioxidants against ultraviolet radiation-induced chronic skin damage in the hairless mouse. Photodermatol Photoimmunol Photomed 1990;7:56–62.
13 Record IR, Dreosti I, Kanstantinopoulos M, Buckeley RA: The influence of topical and system vitamin E on ultraviolet light-induced skin damage in hairless mice. Nutr Cancer 1991;16:219–225.
14 Trevithick JR, Shum DT, Redae S, Mitton KP, Norley C, Karlik SJ, Groom AC, Schmidt EE: Reduction of sunburn damage to skin by topical application of vitamin E acetate following exposure to ultraviolet B radiation: Effect of delaying application or of reducing concentration of vitamin E acetate applied. Scan Microsc 1993;7:1269–1281.
15 Norkus EP, Bryce GF, Bhagavan HN: Uptake and bioconversion of α-tocopheryl acetate to α-tocopherol in skin of hairless mice. Photochem Photobiol 1993;57:613–615.

16 Beijersbergen van Henegouwen GMJ, Junginger HE, deVries H: Hydrolysis of RRR-α-tocopheryl acetate (vitamin E acetate) in the skin and its UV protecting activity (an in vivo study with the rat). J Photochem Photobiol 1995;29:45–51.
17 Kramer-Stickland K, Liebler DC: Effect of UVB on hydrolysis α-tocopherol acetate to α-tocopherol in mouse skin. J Invest Dermatol 1998;111:302–307.
18 McVean M, Liebler DC: Inhibition of UVB induced DNA photodamage in mouse epidermis by topically applied α-tocopherol. Carcinogenesis 1997;18:1617–1622.
19 Jurkiewicz BA, Bisset DL, Buettner GR: Effect of topically applied tocopherol on ultraviolet radiation-mediated free radical damage in skin. J Invest Dermatol 1995;104:484–488.
20 Trevithick JR, Mitton KP: Uptake of vitamin E succinate by the skin, conversion to free vitamin E, and transport to internal organs. Biochem Mol Biol Int 1999;47:509–518.
21 Cesarini JP, Miska P, Poelman M: The effect of antioxidants on human's erythema and sunburn cells. Photochem Photobiol 1988;47:73S.
22 Montenegro L, Bonina F, Ciogilli S, Sirigu S: Protective effect evaluation of free radical scavengers on UVB induced human cutaneous erythema by skin reflectance spectrophotometry. IFSCC Venezia, 1994, pp 769–780.
23 Alberts DS, Goldman R, Xu M-J, Dorr RT, Quinn J, Welch K, Guillen-Rodriguez J, Aickin M, Peng Y-M, Loescher L, Gensler H: Disposition and metabolism of topically administered α-tocopherol acetate: A common ingredient of commercially available sunscreens and cosmetics. Nutr Cancer 1996;26:193–201.
24 Mosmann T: Rapid colorimetric assay for cellular growth and survival: Application to proliferation and cytotoxicity assays. J Immunol Methods 1983;65:55–63.
25 Bando H, Mohri S, Yamashita F, Takakura Y, Hashida T: Effects of skin metabolism on percutaneous penetration of lipophillic drugs. J Pharm Sci 1996;86:759–761.
26 Cannon CL, Neal PJ, Kubilus J, Swartendruber DC, Squier CA, Wertz PW: Lipid characterization of a new epidermal model for skin irritancy and penetration studies. World Congr Alternatives Annu Use Life Sci, Baltimore, Nov 1993.
27 Kriwet K, Parenteau NL: In vitro skin models. Cosmet Toiletries 1996;111:93–101.
28 Data Sheet. MatTek Corporation, Ashland, Mass (USA).
29 Nabi Z, Tavakkol A, Soliman N, Polefka TG: Bioconversion of tocopheryl acetate to tocopherol in human skin: Use of human skin organ culture models (abstract). J Invest Dermatol 1998;110: 1239.
30 Tauber U, Rost KL: Esterase activity of the skin including species variations. Pharmacol Skin 1987; 1:170–183.

Zeenat Nabi, Colgate-Palmolive Co., 909 River Road, Piscataway, NJ 08854 (USA)
Tel. +1 732 878 7103, Fax +1 732 878 7867, E-Mail zeenat_nabi@colpal.com

Author Index

Subject Index